Routledge Revivals

Minority Perspecti

Originally published in 1972, *Minority Perspectives* is the second in a series exploring metropolitan problems within the government structure. The 1960's were a period of civil rights movements as well as poverty in the United States and in the 70's, it became clear that poverty was closely linked to race. This report sets out to explore issues contributing to the metropolitan-minority poverty problem such as racial exclusion and public policy. The papers included in this report discuss issues such as political power in metropolitan areas, the impact an address can have on economic opportunity for minority groups and the effects that laws and litigation can have on poverty. This title will be of interest to students of environmental and urban studies.

Minority Perspectives

Dale Rogers Marshall, Bernard Friedem
and Daniel Wm. Fessler

RFF PRESS
RESOURCES FOR THE FUTURE

First published in 1972
by Resources for the Future, Inc.

This edition first published in 2016 by Routledge
2 Park Square, Milton Park, Abingdon, Oxon, OX14 4RN
and by Routledge
711 Third Avenue, New York, NY 10017

Routledge is an imprint of the Taylor & Francis Group, an informa business

Publisher's Note
The publisher has gone to great lengths to ensure the quality of this reprint but
points out that some imperfections in the original copies may be apparent.

Disclaimer
The publisher has made every effort to trace copyright holders and welcomes
correspondence from those they have been unable to contact.

The publishers would like to make it clear that the views and opinions
expressed, and language used in the book are the author's own and a reflection
of the times in which it was published. No offence is intended in this edition.

A Library of Congress record exists under LC control number: 78186473

ISBN 13: 978-1-138-12119-5 (hbk)
ISBN 13: 978-1-315-65114-9 (ebk)
ISBN 13: 978-1-138-12121-8 (pbk)

Minority Perspectives

NO. 2 IN A SERIES ON

The Governance of Metropolitan Regions

LOWDON WINGO, SERIES EDITOR

Distributed by

The Johns Hopkins University Press, Baltimore and London

Minority Perspectives

Papers by

DALE ROGERS MARSHALL

BERNARD FRIEDEN

DANIEL Wm. FESSLER

Published by Resources for the Future, Inc.

Resources for the Future is a nonprofit corporation for research and education in the development, conservation, and use of natural resources and the improvement of the quality of the environment. It was established in 1952 with the cooperation of the Ford Foundation. Part of the work of Resources for the Future is carried out by its resident staff; part is supported by grants to universities and other nonprofit organizations. Unless otherwise stated, interpretations and conclusions in RFF publications are those of the authors; the organization takes responsibility for the selection of significant subjects for study, the competence of the researchers, and their freedom of inquiry.

This study is the second in a series of papers resulting from an RFF-sponsored project conducted by an informal Commission on the Governance of Metropolitan Regions, chaired by Charles M. Haar of the Harvard Law School. Lowdon Wingo is director of RFF's program of regional and urban studies and a member of the commission. The papers were edited by Jane Lecht.

RFF editors: Henry Jarrett, Vera W. Dodds, Nora E. Roots, Tadd Fisher

Contents

Commission on the Governance of Metropolitan Regions

Foreword

The papers in this volume and its companions are products of a long-standing interest of Resources for the Future in the welfare and development of metropolitan America. More particularly, they stem from an RFF-sponsored project that was launched in the spring of 1970 with the convening in Washington of an informal Commission on the Governance of Metropolitan Regions. Chaired by Charles M. Haar of the Harvard Law School, it is composed of scholars, practitioners, and experienced observers of the metropolitan scene.

"Governance" in the title—The Governance of Metropolitan Regions—is meant to imply something more than government. Webster defines it as "conduct, management, or behavior; manner of life" in addition to "method or system of government or regulation." In these papers, and those that will follow, the authors are concerned not only with the apparatus and process of government in the ordinary sense but also with the total interaction among people in their public capacities and interests, and between people and the public institutions. The dominating question is: How can the governance of metropolis be improved? And next: What must we learn to achieve this? Unless early progress is made in these directions the danger that hard-pressed American cities will crack under the multiple strains of old and new problems will be very real.

RFF did not embark on this effort expecting that metropolitan political reorganization would solve all metropolitan problems, but we are inclined to think that it will help, because we have seen so many obvious steps frustrated by the way in which history has organized our urban political life. Although the reform of the institutions of metropolitan governance is hardly a sufficient condition for the solution of major urban problems, our intuition is strong that it is a necessary one.

This, then, is the theme of the RFF project—the interrelationship of metropolitan problems and governmental structure. It has formed the basis of the deliberations of the Commission and has guided the preparation of the exploratory papers published in this series. The papers do not exhaust the issues of metropolitanism; their purpose is to add some dimensions to an already rich literature, some options for the policy makers. No blueprint for the future is presented, no definitive list of recommendations; the results hoped for are breadth of view, depth of perception at several critical places, and illumination of practical alternatives for action.

Many people contributed to this effort. Charles M. Haar, as chairman, negotiated the contribution of papers and materials. Lowdon Wingo first proposed the Commission as an effective exploratory device, administered the program for RFF, and oversaw the publication of these volumes. Daniel Wm. Fessler of the University of California Law School at Davis has been a valued advisor throughout. Professors Daniel M. Holland of M.I.T. and Karl Deutsch of Harvard made valuable contributions from their time and experience. Michael F. Brewer, while Vice President of RFF, was a faithful and effective participant in the enterprise from its inception. Add to these almost twenty authors and an equal number of Commission members and you have the elements for some new insights to the metropolitan problem. Some are to be found in these papers; more are to be hoped for from succeeding phases of the project.

Joseph L. Fisher
President, Resources for the Future, Inc.

Introduction: Some Public Economics of Social Exclusion

LOWDON WINGO

The pressing domestic problems of the United States in the 1960's–civil rights and poverty–were not uniquely associated with the growth of its metropolitan areas. The civil rights issue was originally construed as a need to repeal the inherited doctrines of racial inferiority of the Negro underlying the culture of the South in order to permit the great mass of black citizens to come into their own under the Constitution. The Supreme Court led the way with a string of decisions that began with *Brown* v. *Board of Education of Topeka* in 1954. The Executive and the Congress followed with a constellation of administrative rulings and legislation that set the national course in a direction of eliminating discrimination against minorities throughout the society.

Poverty, on the other hand, was first seen to be a national phenomenon of segmental failure of the market system, of distressed areas and lagging regions that somehow fell behind in the competition for share in the growth of the national economy. The symptoms were rising unemployment, declining per capita incomes, net outmigration of key elements of local labor forces, and net disinvestment in public and private capital stocks. Accordingly, in the early 1960's the President and Congress addressed themselves to remedies for distressed areas: construction of new infrastructure, easy credit for new enterprise in lagging areas, and increased allocation of public funds to "human capital development" in the form of public health, education, and manpower development programs. Present legislative experience suggests that these policies have become permanent elements of national internal development strategy.

In 1965 in a hitherto unknown locality in Los Angeles an explosion of racial fury and frustration demolished the conventional wisdom about race and poverty as being separate problems "somewhere out there." In the next two years Detroit, Washington, Chicago, New York, Newark, and Hartford were among the cities swept by riots and racial violence. Race and poverty and the rising aspirations of the poor and the black emerged as the dominant social phenomenon of most of the great metropolitan regions of the United States: whereas violence and discrimination in a rural county in Alabama and abject poverty in a mountain town in Eastern Kentucky offended the national conscience, the collapse of orderly social processes in the major urban centers of the nation constituted a threat to the viability of the national society. It was obvious from the beginning that President Johnson's "war on poverty" would have to be fought out along not one but two fronts: the depressed rural areas of the nation, and especially in the South; and (more to the point) the central cities of almost every major metropolitan area in the country. In the rural areas the fundamental issue was how to provide obsolete human resources with a new economic relevance to replace that lost with the technological revolution in the extractive industries. In the cities, the poor were more often than not black, and the black were more often than not poor. The relative frequency of black poverty could only be explained by the persistence of barriers against blacks that not only pervade the society as a whole but have special significance in the institutional organization of metropolitan society.

The peculiarities of the metropolitan-minority poverty problem have arisen from two sets of socioeconomic processes in the prewar period. The postwar transformation of American agriculture resulted in a loss of economic opportunity for literally millions of rural people, many of them southern Negroes, at a time when the major metropolitan regions were the major growth points in the postwar economy. The "push" of capital displacement of labor from extractive industry plus the "pull" of the growth sectors of the economy concentrated in the major metropolitan regions of this country effected a social revolution.

In the 1960–1970 period alone, a net of a million and a half blacks (almost 20 percent of their 1960 population) fled the South: the growth of black populations in New York and California accounted for 80 percent of this outflow, most of the rest going to Illinois, Michigan, and Maryland. In fact, this migration out of the rural economy of the South into Chicago, Cleveland, Cincinnati, Washington, New York, Philadelphia, Boston—as well as to Los Angeles and the Bay area—has transformed the "typical" American Negro from the impoverished and oppressed rural worker of prewar America into an urbanite struggling for a niche in the complex economic and social environment of the city.

The destination of the migrant was the central city in the metropolitan areas where could be found an older housing stock adaptable to the requirements of the poor and which at the same time was the area of maximum access to casual and low-skill employment opportunities. Characteristic of the early stage of this process was the concentrated, densely occupied, black ghetto, which soon overflowed and, under the pressure of continuing inmigration and high rates of natural increase, inundated the older housing stock of the central cities. In Washington the black population advanced from 55 percent to 75 percent of the total population in the ten years ending in 1970, in Baltimore from a third to almost half, Chicago from 15 percent to 22 percent, St. Louis from 30 percent to 40 percent, New York from 15 percent to 22 percent, and Philadelphia from a fourth to a third. Although migration accounted for a good part of the relative change, in almost every major city growth by natural increase equaled or exceeded the figures for net inmigration of minority groups. By 1970 minority groups commanded respectful political attention in the central cities of every major metropolitan region in the country. These postwar trends seem likely to continue.

Interregional migration responding to the great economic changes in postwar America set the scale of the race and poverty issues in metropolitan areas, but the character of intrametropolitan mobility in conjunction with the fragmented structure of government in metropolitan areas gave the problem its unique character. As new highways opened up new supplies of residential land and national housing policies subsidized the new, freestanding, single-family dwelling, white middle-class families progressively abandoned the central city and transferred their citizenship and allegiance to newer, more homogeneous communities growing up in the hinterland. They took with them their specific preferences for public services as well as the resources to support them. Many of them were buying another kind of collective good—freedom from contact with those poorer, darker, or different. They were "voting with their feet," and they knew exactly what they were voting for.

Federally sponsored suburbanization was an amazingly efficient social filter: to qualify for the "suburban club" one had to be willing and able to buy a house (which required a credit rating and qualification by FHA or mortgage-bank housing cost-income formulae); one had to be willing to live like one's neighbors; and one had to have a secure income from a job to which one could afford to travel every working day. Clearly, to be poor was a disqualification and, frequently, simply to be black could be a de facto, if not a de jure, disqualification. Using the machinery of collective choice at its disposal—municipal government—the exclusionary drive was assisted by zoning ordinances, building codes, housing codes, prohibitions against mobile homes, and such other statutory devices as imaginative, middle-class legal talent could devise.

In the first instance what was most violated in these defensive institutional arrangements surrounding the urban middle class was the right of the individual to choose where he would live and with whom he would associate. If it were only that, it would be a grievous loss in a free society, but it turned out to be more. As the white middle class moved out, so did more and more firms, especially those using modern technology or depending on highly technical or professional inputs, or finding access to the rich suburban markets profitable. This relative disinvestment from the central city has eroded central-city tax bases, while at the same time the rising costs of poverty-associated services resulted in upward pressure on property tax rates and another round of disinvestment in the central city. This spiral can only lead to an increasing burden on the central-city resident to pay for deteriorating public services. A second-generation consequence of suburbanization, then, is a growing disparity in the costs and benefits of public services serving the poor and the minorities in the central city as compared with those found in middle-class suburbia.

This migration of capital from the central city to the suburban communities has taxed the poor in other ways also, for jobs have followed investment. Residence in the central city originally had the crucial advantage of immediacy of access to the low-skilled job market, but this advantage began to dissipate as employers moved toward open land beyond economic access of many of the central-city poor. When the low-skilled jobs could not be filled in their new locations, equipment and plant design were adopted to avoid such requirements, and the jobs simply disappeared from the labor markets. Thus suburbanization of business enterprise acted like a wage tax on the poor, which could take any of several forms: higher costs of the journey to work; acceptance of jobs—and pay—beneath one's capabilities; longer periods of unemployment while seeking appropriate jobs; and higher rates of unemployment.

In short, while the social benefits realized by the middle class through exclusionary suburbanization were paid for by the central-city poor whose choice of living environments was narrowed, it was the suburbanization of investment that brought about the metropolitan crisis. The concentration of these minorities with their special needs for welfare and supporting services in economic reservations that have no access to the fiscal resources of the broader region, the diminishing capability for the efficient employment of the human resources so confined, and the refusal of the suburban middle class to assume a broader responsibility for regional well-being—these are the basic ingredients of the metropolitan social crisis of the 1970's.

The late Charles Tiebout observed that the decision to change one's abode within the metropolitan region is in reality a consumption decision by the consumer seeking a more satisfactory configuration of public services *cum* costs, which include not only the taxes he pays but also property-value differentials, incremental costs of private goods, and increments to the costs of the journey-to-work. Ideally, the rational consumer would weigh some summation of these costs against the perceived benefits of different compositions of public services plus those, it must be added, of valued "collective goods" inherent in each locational alternative. Such a good can be defined as a valued quality of a particular residential environment that can be enjoyed by one without decreasing its value for others and that can be sustained by governmental acts or by the allocation of public resources.

For many suburban middle-class families, a valued collective good is freedom from associating with "undesirable" (read "different") social groups. In the metropolitan land market, pursuit of this particular good has two perverse social effects. First, competition for the limited opportunities to locate in such homogeneous communities must bid up location rents, so that people *do* pay more for such housing. Indeed, the prospect of one of the excluded group moving into the community has the effect of destroying this good, so that the much complained-of loss of property values resulting from successful penetration of homogeneous neighborhoods by minority groups is probably a very real economic phenomenon. Resistance to desegregation is clearly based on a real, not imaginary, prospect of a loss of personal welfare. Second, just as there are private goods, such as heroin and unsafe vehicles, whose use we consider to be social vices, so some collective goods can properly be considered to be vicious and without standing in the market or in law in the larger society. This is not because they have harmful consequences for the consumer (although many would have it that they do) but because their enjoyment generates what the economist lovingly calls "negative externalities," that is, they can be enjoyed by one group only by imposing inconvenience, hardship, disadvantage on another group.

The white, middle-class suburbanite will claim a right to his exclusiveness ("likes are drawn to likes") because such tastes are beyond society's arbitration when they affect no one else. But the fact is that exclusionary barriers *do* affect others in quite tangible ways. Although few blacks may consider themselves to be losers if they cannot live next door to white exclusionist Mr. A, a larger number of Mr. A's constrains the supply of locational opportunities available to minorities, a fact that must inevitably force up location rents among the consequently restricted supply of housing opportunities for them. These market processes in part explain both the incredibly high rents and the appalling quality of housing available to minority groups in the central cities of almost every major metropolitan area in the United States. Hence, the collective good of a homogeneous, middle-class suburbia is paid for by the higher property values facing the suburbanite home buyer and by the resulting tax on the minority tenant of the central city in the form of a substantial increment to rents and a substantial loss of housing quality. Quite obviously, exclusionary association of the middle class is a socially vicious good, which has powerful regressive income consequences for the poor and the excluded minorities.

A different case is sometimes made by the proponents of homogeneous communities. If the metropolitan area were organized into a substantial number of homogeneous communities, each producing its own public goods, it is much more likely that more people will get more of what they want from the resulting output of public services than if the area were organized into the same number of heterogeneous communities. Such a proposition, however, can be maintained only if all parties have equal freedom of association and equal access to the region's fiscal resources, conditions manifestly inconsistent with the current reality of metropolitan political economy. Homogeneity is enjoyed by the white middle-class suburbanite with high quality services and *relatively* low tax burdens; it is endured by the central-city minorities with deteriorating public services and high tax burdens.

Since the resulting crisis is compounded of (1) the obsessive preoccupation of the suburbanite with the collective good of social homogeneity and (2) of the institutional arrangements he has been able to exploit in the passionate pursuit of exclusion, what are the policy alternatives to the crisis? John Kain has labelled one "the gilded ghetto," the acceptance of a dualistic metropolitan society with compensatory public policies to upgrade rapidly the economic status of the excluded minorities. Those who value homogeneity are left to enjoy it, while those who dislike it are compensated by the use of public resources to improve their own environments so that they do not bear the costs of the enjoyment of homogeneity by others (assuming, of course, that the compensation is something more than token). The costs would then be borne by the non-central-city dweller, with the possibility that the suburbanite's homogeneity would be subsidized by nonmetropolitan America instead.

The alternative strategy posits that homogeneity dependent on exclusionary practices of the public agent in any of its forms is as vicious a product in this society as heroin or criminal behavior. This assumption, however, does not solve the problem. To do that, we have either to change the hearts of men (or at least their addiction to the social vice of publicly mandated homogeneity), a spiritual rather than a political operation, or to deprive the exclusion-bent of their access to any instruments or comfort of public institutions, a less remote possibility, perhaps, and one which comes to the forefront with the prospects for the reform of the governmental institutions of metropolitan regions.

This small volume does not set out to give any thoroughgoing evaluation of these alternatives; it simply seeks to examine some of the real and inferred consequences for the poor and the minority citizens of metropolitan regions. It is not at all obvious that metropolitan government reform will carry with it de facto improvement of the conditions and standing of these groups within metropolitan society, but it is clear that no strategy of reform can be decided upon without a careful and convincing appraisal of its consequences for the groups most exposed to damage. In many ways reform is a middle-class concept addressed to the realization of middle-class values; to contribute to the resolution of the metropolitan social problem of the 1970's, it must be something else. It must enfranchise minorities with equal access to the machinery of public policy and the levers of public decision making. The voices of doubt about this are loud and insistent and must be answered in the design of metropolitan reform.

Dale Rogers Marshall, in the essay that follows, appraises what is known about the impact of metropolitan consolidation on the political standing and economic environment of minority groups where reform efforts, successful and unsuccessful, have become major metropolitan political issues. It comes as no surprise that too little is known to be conclusive about how minorities are likely to fare under reform of metropolitan governance. Nor is it surprising that minority political power frequently becomes diluted in larger units of government; here analysis heightens rather than sets to rest apprehensions that minority groups may indeed have much to lose with some kinds of governmental change.

Bernard Frieden, in contrast, examines the lot of the minority groups who do escape to the suburbs. What are the "costs" of intrametropolitan integration? Is the lot of the black suburbanite substantially improved by a change of address? How does he fare in the labor market? Can he get the public services he needs in communities where the composition of public services is determined by his white neighbors? Does his voice carry more weight in the things that affect his life? There is some reason to discount the rosier estimates of liberal integrationists, he argues, but this may simply make the case for a more carefully considered, more elaborate public strategy for integration.

Finally, Daniel Wm. Fessler argues that the law, as it comes to bear on the inequities that governments and tradition have generated, is changing at a revolutionary pace, but that organizational structures are not. New force is being placed behind the demands of minorities for a fair share in the distribution of the societal goods for which they have been oppressively taxed in the past. The reform of metropolitan governance will have to take place simply to embody the legal reforms now taking place as the courts relate the Constitution to twentieth century metropolitan society.

We know a great deal about the historical process that brought the ethnic minorities from their rural origins to present metropolitan homes. We understand a great deal about the social ecology that has increasingly made the central city a beleaguered bastion of minority aspiration and the suburbs defensive fortifications against the erosion of their middle-class social homogeneity. We have observed the employment of government institutions in bringing about this critical social state. Nevertheless, the continuing debate on the reform of metropolitan governance is rarely constrained by the need to resolve this impasse. Such a resolution can only be brought about by institutional restraints on the privileges hitherto enjoyed by some of imposing the costs of their life styles on others. This is perhaps a more fruitful place to begin the design of metropolitan reform than with questions of cost savings from public services. Everybody knows who will bank those savings.

1 Metropolitan Government: Views of Minorities

DALE ROGERS MARSHALL*

Proposals for metropolitan government are not new. For at least forty years, analysts have been discussing the fragmentation of local governments in the United States and calling for metropolitan reorganization to solve the social, political, and economic problems in our urban centers,[1] but the approaches to metropolitan government have changed over the years.

The early approach assumed that fragmentation was a problem and prescribed the integration of all local government units into a unified metropolitan government. New approaches developed in the 1950's, due to the phenomenally consistent failure of reform proposals in elections and to behavioral developments within social science. One approach analyzes the politics of governmental reorganization attempts, showing the forces influencing the outcomes.[2] Other approaches start by looking closely at the

*Lecturer, Department of Political Science, University of California, Berkeley.

[1] See, for example, Paul Studenski, *The Government of Metropolitan Areas in the United States* (New York: National Municipal League, 1930); R. D. McKenzie, "The Rise of Metropolitan Communities," in *Recent Social Trends in the United States* (New York: McGraw-Hill, 1933); Victor Jones, *Metropolitan Government* (Chicago: University of Chicago Press, 1942).

[2] See Robert Warren's description of new approaches in "Political Form and Metropolitan Reform," *Public Administration Review*, September 1964. Books concerned with the politics of reform include David Booth, *Metropolitics: The Nashville Consolidation* (East Lansing: Michigan State University, 1963); Daniel Elazar, *A Case Study of Failure in Attempted Metropolitan Integration: Nashville and Davidson County, Tennessee* (Chicago: National Opinion Research Center, 1961); Scott Greer, *Metropolitics: A Study of Political Culture* (New York: Wiley and Sons, 1963); Henry Schmandt et al., *Metropolitan Reform in St. Louis: A Case Study* (New York: Holt, Rinehart and Winston, 1961); Edward Sofen, *The Miami Metropolitan Experiment* (Bloomington: Indiana University Press, 1966).

actual workings of fragmented government to see what, if any, problems result. Some of these conclude that structural reform is not needed; others criticize the adequacy of proposals for unified metropolitan government and suggest alternative reform of metropolitan structure.[3]

Yet all these treatments of metropolitan reform have at least one trait in common—they have not focused on its impact on minorities. The actual or projected effects of reorganization on blacks, browns, Indians, and Puerto Ricans are considered superficially, if at all. Individual works sometimes include subsections on the way Negroes have voted on reform proposals and their attitudes toward reform, but most of the space is typically devoted to discussions of implications for suburbs and central cities.[4] Of course the term central city often encompasses minorities, but it includes many other interests such as downtown businessmen and white residents. So the racial and ethnic aspects of reorganization are too often implicit rather than explicit. Minority interests are given much less attention than majority interests.

As we have become aware that race and ethnic issues are central to a wide variety of urban problems, reorganization must be expressly reexamined in these terms. Since metropolitan reform involves changes in the divisions of power and rewards, the stakes for minorities are potentially very high.[5] As one step in the needed reordering of priorities, this paper reviews the literature relevant to minorities' relationship to metropolitan reorganization. After a brief review of the trend in urban reform, consideration is given first to the attitude of minorities to metropolitan government and then to the evidence on the effect of such government on minorities.

The Trend in Metropolitan Reform

The existence of myriad governmental units in metropolitan areas has been described as "the local governmental jungle."[6] The boundaries and jurisdic-

---

[3] See Vincent Ostrom, Charles Tiebout, and Robert Warren, "The Organization of Government in Metropolitan Areas," *American Political Science Review*, December 1961, pp. 831–42; Edward Banfield, "The Politics of Metropolitan Reorganization," *Midwest Journal of Political Science*, May, 1957, pp. 77–91; Robert O. Warren, *Government in Metropolitan Regions: A Reappraisal of Fractionated Political Organization* (Davis: Institute of Governmental Affairs, University of California, 1965); John Bollens and Henry Schmandt, *The Metropolis: Its People, Politics, and Economic Life* (New York: Harper and Row, 1970).

[4] Summarized in Bollens and Schmandt, *Metropolis*, Chapter 14; see also Willis Hawley, "Toward a Theory of Metropolitan Political Integration" (mimeographed, January 1969) and compare the limited citations on minorities with the fuller citations on suburbs, classes, and whites.

[5] Greer, *Metropolitics*, p. 1.

[6] United States Advisory Commission on Intergovernmental Relations, *Urban America and the Federal System* (Washington, D.C., October 1969).

tions of cities, counties, school districts, and special districts typically overlap and conflict in a complex maze. For example, the Chicago Standard Metropolitan Statistical Area (SMSA) has over one thousand local governments; Philadelphia has over eight hundred.[7] The interdependent activities of people in the metropolis cannot be dealt with comprehensively by any one government. The result is said to be inferior services, economic inefficiencies and disparities, and lack of responsiveness to popular control. Creation of larger governmental units is seen as a way to end fragmentation and the problems of service, economics, and control thought to be associated with it.[8]

The most obvious method of creating larger governmental units involves radical reform. Voters are asked to approve legal changes that create different types of metropolitan governments. The purest type consists of only one level of government and requires consolidation of existing governments.[9] But the two-level, limited type of metropolitan government has become more popular, allocating certain functions to an areawide unit and retaining certain functions for existing units. This reform can be accomplished by the creation of metropolitan multipurpose districts, urban counties, or federations.[10]

Even though direct radical reform attempts are the most widely known means of creating metropolitan government, indirect incremental reforms have actually become more significant.[11] Although most radical reforms have been failing to receive voter approval, incremental reforms have been responsible for the clear trend toward larger governmental units for metropolitan areas. This trend has resulted from both local initiative and federal stimulation. Examples of the trend are numerous. Formal and informal interjurisdictional cooperation has increased in urban centers as evidenced by the growth

[7] Bollens and Schmandt, *Metropolis*, p. 102.

[8] Scott Greer, *Governing the Metropolis* (New York: Wiley and Sons, 1962), Chapter 4; James Fesler, ed., *The 50 States and Their Local Governments* (New York: Alfred Knopf, 1967), p. 537; United States Advisory Commission on Intergovernmental Relations, *Fiscal Balance in the American Federal System* (Washington, D.C., October 1967).

[9] See Bollens and Schmandt, *Metropolis*, Chapter 14, which cites Nashville 1962 and Jacksonville 1967 as the most recent examples of exceptions to the pattern of voter rejection such as occurred in St. Louis.

[10] Bollens and Schmandt, *Metropolis*, Chapter 12, discusses Toronto's federation and Miami's urban county, as well as the failures to create urban counties in Cleveland, Dayton, Houston, and Pittsburgh.

[11] Peter Marris and Martin Rein, *Dilemmas of Social Reform: Poverty and Community Action in the United States* (Chicago: Atherton Press, 1967), p. 15, points out that in the 1950's the Ford Foundation saw the direct creation of metropolitan government as a promising approach for solving urban problems but became disillusioned and changed to other strategies. The incremental strategies have aptly been called the "quiet revolution." See the address by Norman Beckman, Director, Office of Intergovernmental Relations and Urban Program Coordination, U.S. Department of Housing and Urban Development, before the 73rd National Conference on Government of the National Municipal League, Milwaukee, Wisconsin, November 13, 1967.

of intercity agreements and councils of government.[12] The number of single-purpose special districts to handle metropolitan problems such as air pollution has grown.[13] And the federal social legislation of the 1960's also helped urban areas bypass the difficulties of formal metropolitan reorganization by encouraging the creation of areawide single-purpose governmental units.[14] Much federal aid to cities was made contingent upon the approval of areawide planning bodies responsible for the coordination of policies and projects in fields such as poverty, health, pollution, and housing.

The Economic Opportunity Act of 1964 is an example of this approach. It made funds available to new entities called community action agencies. The boards of these agencies, designed to be broadly representative of the areas served, are responsible for screening and funding applications for poverty money coming from their jurisdiction.[15] Comprehensive health planning councils are the comparable units encouraged by federal legislation in the health field; and councils of government often fulfill this function in planning as provided for in the Housing and Urban Development Act of 1965 and the Demonstration Cities and Metropolitan Development Act of 1966.[16]

The trend toward larger governmental units in the form of areawide, single-purpose governments has been criticized for increasing the fragmentation of metropolitan policy making (functional rather than geographical fragmentation). Such governments are said to prevent comprehensive consideration of priorities; for example, a water quality control board orders a city to spend large sums of money on sewage treatment without having to face the question of whether there are other more important claims on that money, such as education. Single-purpose governments are also criticized for being removed from public scrutiny; they are the domain of professionals, or functional specialists. They contribute to what has been called "the vertical functional autocracy."[17]

Dissatisfaction with single-purpose metropolitan units has led to interest in the creation of multipurpose metropolitan units as the next step in the continuing trend to larger governmental units. The Intergovernmental Coopera-

---

[12] Bollens and Schmandt, *Metropolis*, Chapter 13; Hawley, "Toward a Theory," p. 4.

[13] Bollens and Schmandt, *Metropolis*, pp. 313–17.

[14] Terms such as "units" and "bodies" are used here as general labels to avoid the complexities involved in more precise legal descriptions, such as "special districts."

[15] Community action agencies do not necessarily serve a whole metropolitan area, but some such as Los Angeles' agency cover a large proportion of a metropolis. Minimum-size requirements were set as an incentive for areawide agencies.

[16] See Alan Altshuler's review of the federal government's encouragement of metropolitan decision making, in *Community Control: The Black Demand for Participation in Large American Cities* (New York: Pegasus, 1970), p. 183.

[17] Bollens and Schmandt, *Metropolis*, pp. 317–20, and ACIR, *Urban America*, p. 5.

tion Act of 1968 is one move in this direction. It says general-purpose units will be given priority for federal aid and requires the planning bodies (often councils of government) designated by the Department of Housing and Urban Development (HUD) to review all applications from local planning and regulatory agencies and local governments, including applications from community action agencies, comprehensive health planning groups, air pollution districts, etc.[18] Thus the planning bodies are given new functions as coordinators of the federally created single-purpose coordinating bodies. Whether the review will be more than an empty formality is not yet clear, but the increase in the regional planning bodies' responsibilities could be a step toward the creation of multipurpose metropolitan governments.[19]

Thus, while citizens and scholars debate the desirability of metropolitan government, the emergence of metropolitan governments is well under way. At the same time there are growing demands for community control, for reversing the trend toward larger governmental units. What is the reaction of minorities to the metropolitan trend? What impact will it have on minorities? These are the questions to which we now turn.

Attitudes of Minorities to Metropolitan Government

Information on the attitudes of minorities to metropolitan reorganization can be pieced together from their votes on radical reform proposals, their reactions to metropolitan governments which come into existence after proposals are passed, and their comments on existing and proposed incremental reforms. Unfortunately, evidence on attitudes of Mexican-Americans, Indians, and Puerto Ricans is almost nonexistent, so this discussion focuses on blacks; but the term "minorities" is used throughout the paper to emphasize the need to find out more about other minorities' views of metropolitan reform. The available information is also marred by the treatment of black opinion as uniform, or at best by distinguishing only between black leaders and all other

---

[18]See Altshuler, *Community Control*, p. 183, and Bureau of the Budget memo A-95.

[19]Stanley Scott reviews other approaches toward the creation of multipurpose metropolitan governments in "The Regional Jobs to be Done and Ways of Getting Them Accomplished," Background Paper No. 2, Regional Conference–1970 (Institute of Governmental Studies, University of California, Berkeley, April 18, 1970). He describes the umbrella approach currently being used in Minneapolis-St. Paul and proposals for the San Francisco area. For a multipurpose approach to environmental quality, see Ruth Roemer *et al.*, "Environmental Health Services: Multiplicity of Jurisdictions and Comprehensive Environmental Management" (mimeographed draft, Institute of Government and Public Affairs, UCLA, June 1970). See also Stanley Scott and John C. Bollens, *Governing a Metropolitan Region: The San Francisco Bay Area* (Berkeley: University of California, Institute of Governmental Studies, 1968).

blacks. More detailed breakdowns are needed so that finer distinctions are possible within the Negro community. Different groups of Negroes perceive various goals and problems to be important. But we know almost nothing about what segments are more favorable to governmental reorganizations and why. What groups currently without opinions on the issue might be mobilized to support reform? Are the apparent attitudes more the voice of black politicians with a stake in the existing election districts than of the rank and file?

The existing information does reveal two main lines of argument about metropolitan government and minorities. The main argument of minorities against any type of metropolitan government is concerned with *power*. They say metropolitan government dilutes the power of minorities which has been increasing in central cities. Some feel this dilution is intentional, that metropolitan reform is designed with such a goal in mind. Pointing to the growing number of minority councilmen and black mayors, minorities say such political representation would be threatened if suburban voters were added to city voters in metropolitan elections. Banfield called attention to this political effect of metropolitan government in 1957, saying power would go from lower-class Negro and Catholic elements to middle-class white and Protestant elements.[20] The second argument is concerned with the quality of *services*, but it is not made very often by minorities.[21] It typically comes from white commentators who profess an interest in minority problems. Metropolitan government is said to promise better services to minorities because suburban money will become available to ease the problems of minorities.[22] The absence of minority defense of metropolitan government on the grounds of improved services, does *not* mean minorities are unconcerned with quality of services. To the contrary, the evidence suggests urban services are a major interest of minorities.[23] For example, a 1966 UCLA survey of Negroes in the Los Angeles riot area emphasizes their grievances about poor neighborhood conditions (such as dirty streets, dilapidated housing) and police protection.

[20] Banfield, "The Politics of Metropolitan Reorganization," p. 87.

[21] Greenstone and Peterson show the utility of distinguishing between power and services in the war on poverty, "Reformers, Machines, and the War on Poverty," in James Wilson, ed., *City Politics and Public Policy* (New York: Wiley and Sons, 1968).

[22] Willis Hawley, "Black America's Stake in the Political Reorganization of Metropolitan Areas" (mimeographed draft, 1970).

[23] The studies referred to in the following are: Raymond Murphy and James Watson, "The Structure of Discontent: The Relationship Between Social Structure, Grievance, and Support for the Los Angeles Riot" (Los Angeles: University of California, Institute of Government and Public Affairs, 1967), pp. 17 and 114; *Supplemental Studies for the National Commission on Civil Disorders*, July 1968, pp. 39–41; see also the *Field Surveys* done for the President's Commission on Law Enforcement and Administration of Justice, I, pp. 119 and 122, and II, p. 58.

Similarly, studies done for the National Commission on Civil Disorders also indicate Negroes are more dissatisfied than whites with city services such as police, parks, and garbage collection. But minorities' intense concern with services is also accompanied by more criticism of governmental efforts to deal with urban problems, so it is not surprising they are skeptical about the ability of even a reorganized government to actually provide better services.

Votes on Radical Reform

A majority of blacks typically opposes metropolitan reorganization proposals as shown in the literature on politics of reform. The studies of Cleveland, St. Louis, Miami, and Nashville, all show Negroes voting against metropolitan reform proposals. However, aggregate voting data do not permit identification of race and the surveys of attitudes toward reform were not designed to delve into racial aspects of the issue, so the data only give incomplete information on the attitudes of blacks in these cities.

Greer in *Metropolitics*[24] describes the opposition of Cleveland Negro political organizations to reorganization. They felt the 1959 urban county charter proposal in Cleveland would prevent the election of more than two Negro representatives and weaken civil service provisions which protected their people. One leader stressed, "My people have to be treated with respect." Instead, charter campaign workers asked the Negro leaders how much their support would cost. Such perceived insults led to active campaigning against the reformed charter in Negro precincts, and the vote was two to one against it, contributing significantly to the defeat.[25] A study of votes on charter reform issues over twenty-five years shows the attitudes of Negro wards became more negative to metropolitan reorganization as the number of predominately Negro wards increased in Cleveland.[26] As power grew, Negroes apparently felt they had more to lose from metropolitan reform.

Greer reports similar opposition by Negro leaders and voters to St. Louis's 1962 city-county consolidation proposal. The leaders refused to support anything that they had not helped formulate.[27] The committees campaigning for the reform avoided contact with the Negro leaders to prevent publicizing black leaders' opposition and increasing the turnout of black negative votes.[28]

[24] Greer, *Metropolitics*, pp. 94-95.

[25] *Ibid.*

[26] Richard Watson and John Romani, "Metropolitan Government for Metropolitan Cleveland: An Analysis of the Voting Record," *Midwest Journal of Political Science* 5 (November 1961): 365-90.

[27] *Metropolitics*, p. 80. Although some mention is made of Negro representation on charter commissions in Cleveland, St. Louis, and Miami, Greer does not give the exact numbers and leaves the impression they were a minor element, p. 25.

[28] *Ibid.*, p. 90.

Evidence from Miami shows Negro precincts opposed the 1957 urban county reorganization by about 60 percent, but since Negroes were only about 7 percent of the registered voters, their negative votes did not prevent the success of the reform proposal.[29] Likewise, negative black votes on Nashville's city-county consolidation reform in 1962 did not defeat it. A majority of Negroes voted against Nashville's metro proposals in both 1962 and 1958, which indicates they accepted their leaders' views that reorganization would weaken Negro voting strength.[30]

Even though the evidence on minority voting on metropolitan reorganization is incomplete, the general pattern seems clear. A majority is not convinced that possible service improvements will adequately compensate for their loss of power, so they oppose radical reform proposals. Turning to consideration of the reaction of minorities to metropolitan governments that have been approved by voters, the scarcity of information makes generalizations more difficult.

Reactions to Reformed Governments

In view of past interest in metropolitan reorganization, the incompleteness of studies on the effect of metropolitan governments in operation is surprising. Grant's articles on the Nashville experience present evidence of positive accomplishments, including clearer lines of responsibility, increased professionalization, elimination of duplications, equalization of services, elimination of the inequitable financial burden on the city, and no loss of accessibility. But the equalization of services meant improved services for the suburbs, not for the city. Services for the central city were not reduced, says Grant, but he does not claim they were improved.[31] The accomplishments of Toronto also are predominately services for the outlying areas rather than

---

[29] Sofen, *Miami Metropolitan Experiment*, p. 78.

[30] Brett Hawkins, *Nashville Metro: The Politics of City-County Consolidation* (Vanderbilt University Press, 1966), p. 132; Hawley, "Black America's Stake," p. 4, reports Nashville blacks favored reform in 1962; T. M. Scott, "Metropolitan Governmental Reorganization Proposals," *Western Political Quarterly* 21 (June 1968): 252–61, reviews eighteen reorganization attempts between 1950 and 1961 and finds major change attempts were defeated except in four cases (Atlanta, Newport News, Miami, and Nashville) where special circumstances existed. See also Bollens and Schmandt, *Metropolis*, Chapters 11, 12, and 14.

[31] Daniel Grant, "A comparison of Predictions and Experience with Nashville 'Metro', " *Urban Affairs Quarterly* 1 (September 1965): 34–54; Daniel Grant, "Political Access under Metropolitan Government: A Comparative Study of Perceptions by Notables," in Robert Daland, ed., *Comparative Urban Research: The Administration and Politics of Cities* (Beverly Hills: Sage Publications, 1969). See also citations in Bollens and Schmandt, *Metropolis*, pp. 305–6.

solutions of social problems of particular concern in the central city.[32] And Sofen's assessment of Miami emphasizes the problems in financing and leadership which have plagued the new government in spite of some increases in efficiency.[33]

But these studies are often impressionistic and do not analyze the extent to which metropolitan government was responsible for the changes. Furthermore, they say almost nothing about the effect of the reforms on minorities and the minorities' attitudes to the reforms.

Grant's study of knowledgeable observers in Toronto, Miami, and Nashville does provide some information on these topics. He interviewed a total of fifty-six leaders about their perceptions of the effect of metropolitan government and found that 79 percent of the respondents in Miami and 70 percent in Nashville felt Negroes had easier access, or at least no decrease in access, because of metropolitan government. In Miami the explanation was that professionals had replaced the old rural-type politicians. In Nashville this access was attributed to the elimination of split responsibilities and the creation of single-member districts designed to insure the election of several Negroes to the metro council.[34] Grant concludes metropolitan government will not inevitably weaken Negro political access. Yet he only interviewed three Negroes among the thirty-nine respondents in Miami and Nashville. Even though they were favorable toward metro, there is no way of knowing to what degree they reflect the views of other Negro leaders or the Negro public. Also one suspects that an increase in access in southern cities is measured by the previous extreme deprivation and may not be due to metropolitan government.

The study of the effect of Nashville's metro does show that Negro voting strength has not been diluted. In no instance, says Grant, has a candidate supported by a large majority of Negroes been defeated by added white votes. Negro voting strength in Nashville continues to be based on their ability to hold the balance of power between white factions, rather than on their ability

---

[32] Harold Kaplan, "Metro Toronto: Forming a Policy-Formation Process," in Edward Banfield, ed., *Urban Government: A Reader in Administration and Politics* (New York: Free Press, 1969); and Harold Kaplan, *Urban Political Systems: A Functional Analysis of Metro Toronto* (New York: Columbia University Press, 1967). See also Bollens and Schmandt, *Metropolis*, pp. 344–46.

[33] Sofen, *Miami Metropolitan Experiment*. See also citations in Bollens and Schmandt, pp. 305–6 and a listing of positive accomplishments, pp. 332–34.

[34] Grant, "Political Access," p. 262. Grant defines access as the ability of groups "to get an attentive hearing from responsible officials when they feel they have a problem." Note his reliance on white opinions of black access and the absence of a question on black power in the sense of the ability to determine policy outcome. For the distinction between power and access, see R. E. Agger *et al.*, *The Rulers and the Ruled* (New York: Wiley and Sons, 1964), pp. 51–58.

to elect by their votes alone. Since metro, when Negroes vote for a losing candidate they are outvoted not only in white suburbs but in the white old city as well.[35]

So we have some studies of white perceptions of metro's effect on Negroes and services in the central city, but we need to look systematically at minority attitudes and services for minorities. Further generalizations about power require analysis of minority voting strength in metro elections and minority representation on metro bodies compared with voting and representation records before metro.

Even though the absence of systematic studies makes conclusions about the attitudes of minorities toward reorganized metropolitan government dangerous, vocal black leaders have not been hesitant in expressing their dissatisfaction with reformed governments. In Nashville charges have been made that housing, highways, and police services are not responsive to minorities and that minority political power has been diluted.[36] Julian Bond of Georgia says black power has been diminished in both Nashville and Miami.[37] Yet this opposition to metropolitan government may not be more intense than current minority criticism of traditionally governed cities.

Nevertheless, such criticism of existing metropolitan governments contributes to the general impression of minority opposition to metropolitan reorganization. When combined with the opposition to reorganization proposals already discussed, minorities appear to have a very negative view toward metropolitan government, but before such a conclusion can be supported, more probing analyses are needed of the attitudes of varying segments of minority communities.

Reactions to Incremental Reforms

What is the attitude of minorities to the various single-purpose and proposed multipurpose governmental units described in the first section of this paper? The attitudes toward incremental reforms creating larger governmental units appear more varied, perhaps because of the diverse types of areawide bodies existing in various cities. One is forced to speculate in the absence of information.

Minorities undoubtedly feel at least as isolated from most single-purpose bodies as the rest of the population does. And since many of these entities deal with physical problems, they apply technical, professional values not noted for their sensitivity to social or minority perspectives; thus, minorities

---

[35] Grant, "A Comparison of Predictions and Experience," p. 52.

[36] Bollens and Schmandt, *Metropolis*, p. 306.

[37] See Armando Rendon, "Metropolitanism: A Minority Report," *Civil Rights Digest* 2 (Winter 1969): 8.

insofar as they are aware of these entities at all may feel they are particularly harmful to their interests. For example, transportation districts may design rapid transit to meet the needs of the affluent but not of the minorities, or redevelopment projects may concentrate on helping downtown businesses. However, some single-purpose metropolitan units, especially those dealing with social problems, have become the province of minorities. The clearest examples are found among community action agencies. Kramer describes this phenomena in San Francisco, where blacks and browns got control of the resources available through community action programs, including jobs, power, and status. The resources gave ethnic groups the motivation to organize to influence this new decision-making forum and their participation gave them "public attention, control over some new social service resources and the patronage connected with them, as well as an increased measure of political sophistication."[38] The same pattern may be occurring in other social agencies that require participation of the clients, such as model cities programs.[39] But apparently minorities have little voice in single-purpose agencies such as health planning councils which draw heavily on white professionals or councils of government dominated by local government officials. However, HUD has been telling councils of government to include minority members.[40] If this requirement increases minority representation on such councils, minorities could begin to look on them with more favor.

The conclusion here is that the attitudes of minorities to single-purpose areawide units vary, depending on the voice they feel they have in each one. Thus their reactions to proposed multipurpose districts will depend partly on the relation of such districts to single-purpose units they favor. For instance, minorities undoubtedly would oppose a multipurpose district which took over the functions of a community action agency controlled by minorities unless they were given equal control of the new entity. A survey of San Francisco area leaders' attitudes to regional government shows some of the additional factors of concern to minority leaders. There were 205 interview respondents, 8 of whom were black or brown. Although they were as likely to endorse the need for a limited, multipurpose regional government as white leaders, they were generally ambivalent and not interested in pushing for its creation. They doubted the new government would advance the interests of

---

[38] Ralph Kramer, *Participation of the Poor* (Englewood Cliffs, N. J.: Prentice-Hall, 1969), p. 249. See also Dale Marshall, "Public Participation and the Politics of Poverty," in Peter Orleans, ed., *Urban Affairs Annual Review* 5 (Beverly Hills: Sage Publications, 1971).

[39] For a review of these programs, see Melvin Mogulof, "Citizen Participation: A Review and Commentary on Federal Policies and Practices," mimeographed working paper, The Urban Institute, Washington, D.C., January 1970.

[40] Stanley Scott, "The Regional Jobs to be Done."

the poor. Unlike the whites, the minority respondents felt regional government had to be concerned with social, and not just physical, problems. The minorities also felt the governing board of any regional government should be elected from single-member districts.[41] Black Assemblyman Willie Brown from San Francisco voiced these sentiments when he said minorities will not support regional government which gives priority to the environment rather than social problems and which does not provide for minority representation by direct election.[42]

Comments by other minority leaders about the incremental trend toward any form of large governmental units also indicate their reservations. Julian Bond is quoted as saying, "if there were no racial considerations, metropolitanism would be desirable, but black people have to fight it until they can get certain guarantees of equitable representation."[43] Dr. Ernesto Galarza says minorities do not know how to deal with existing governmental agencies and metropolitan governments will just add another layer of confusion. This layer will hinder their attempts to make themselves heard, but most of them are not aware of these dangers, he says. All the devices, says Galarza, "being discussed in order to make the metropolis more manageable are not addressing themselves to minority groups."[44]

To summarize, the limited evidence on the attitudes of minority leaders and voters toward metropolitan radical reform proposals, existing metropolitan governments, and proposed incremental reforms indicates opposition to metropolitan government of any type. These people feel their power will be diminished and the possibility of improved services for minorities will not be real, since the metropolitan governments will not be responsive to minority interests. Their predominant opinion is "in the sometimes purposeful, but most often willy-nilly shifting of governmental centers, of redistribution of resources and of political power plays, the man of color or cultural differences tends to be the loser."[45] None of the forms of metropolitan government then is seen as offering satisfactory solutions to the metropolitan problems of minorities.

Yet, as previously mentioned, this apparently uniform opposition may be misleading. First, we do not have enough information on the attitudes of

---

[41]Willis Hawley, "The Future of Regional Government in the San Francisco Bay Area: A Study of Leadership Attitudes," Institute of Governmental Studies, University of California, Berkeley (typed draft, December 1969).

[42]Willie Brown, Jr., "Regional Government: Impact on the Poor," in Harriet Nathan and Stanley Scott, eds., *Toward a Bay Area Regional Organization* (Berkeley: Institute of Governmental Studies, 1969).

[43]Rendon, "Metropolitanism," p. 9.

[44]*Ibid.*, p. 10.

[45]*Ibid.*, p. 6.

diverse groups. Second, the opposition is undoubtedly part of a suspicious-ness of governmental institutions in general, not just of metropolitan govern-ment. For example, black leaders have spoken out against federal emphasis on environmental pollution because they feel it results in a lower priority for jobs and housing. So criticism of a similar emphasis at the metropolitan level is not surprising. But the significance of criticism of metro forms arises be-cause metropolitan government is a newly emerging governmental institution and the struggle over the division of power and rewards has not yet been resolved.

Evidence on the Effects
of Metropolitan Government on Minorities

Having reviewed the negative attitudes of minorities to metropolitan govern-ment, the question arises as to how justified these feelings are. What social science evidence is relevant to the effects of metropolitan government on minorities?

The disagreement among social scientists on the desirability of metropoli-tan government has already been indicated. They also differ on the effect of reorganization on minorities. Piven and Cloward think it will be harmful and is being advocated more intensely now as part of a federal strategy to deny minorities political power in the central city and thus appeal to white subur-ban voters.[46] Altshuler says whites become more favorable to metropolitan government as black political power grows. (One wonders if fear of black control of central cities will overcome northern white suburban commitment to local home rule and isolation from central city problems.)[47] For totally different reasons, Banfield would also deny that metro could ease minority problems. His *Unheavenly City* is devoted to urban social problems without any discussion of metropolitan government, but Banfield's implication is clear—structural reforms will not bring about any solutions to human prob-lems for minorities or anybody else.[48]

What evidence is there that change in structure *can* improve services for minorities? Little support currently exists for the naive argument that elimi-nation of fragmentation alone will upgrade minority services. The service

---

[46] Frances Piven and Richard Cloward, "Black Control of Cities," *New Republic*, September 30, 1967 and October 7, 1970.

[47] Altshuler, *Community Control*, p. 51; Banfield, "Politics of Metropolitan Re-organization," p. 89. For a discussion of the way small municipalities protect the values of the inhabitants see Oliver Williams, "Life Style Values and Political Decentralization in Metro Areas," in Terry Clark, ed., *Community Structure and Decision-making: Com-parative Analyses* (San Francisco: Chandler Publishing Co., 1968).

[48] Edward Banfield, *The Unheavenly City* (Boston: Little, Brown, 1970).

problems are not created by fragmentation so they cannot be solved merely by metropolitan government:[49] local communities fail to deal with many of the problems "not for lack of area-wide planning, but for lack of political will."[50] Governmental structure alone guarantees nothing because problems will still exist due to incompetent leadership and the unwillingness of people to have taxes raised.[51] Nevertheless, some social scientists believe metropolitan government will permit "many things that would otherwise be impossible or a great deal more difficult."[52] The capacity to act is as essential as the will to act, and without metropolitan government, that capacity does not always exist, they argue.[53]

In support of this view, attempts have been made to determine which problems could best be handled by areawide, as opposed to local, governments. In what functional areas is capacity for solving problems most seriously damaged by fragmentation? Economic, political, and administrative criteria have been suggested for making this determination.[54] Economic criteria include consideration of the size necessary for economies of scale to occur in a given function. But even the evidence from the application of such an apparently quantifiable standard is mixed because different costs and indicators are used in the calculations.[55] Yet some conclude that economies of scale are possible and also economies due to elimination of overlapping and increase in specialization. And while proponents of metro's ability to improve services no longer promise a decrease in total expenditures (Nashville and Toronto expenditures went up due to inflation and increased demand), they

---

[49] Paul Friesma, "The Metropolis and the Maze of Local Government" in Scott Greer, *et al., The New Urbanization* (New York: St. Martin's Press, 1968).

[50] Piven and Cloward, "Black Control," September 30, 1967, p. 21.

[51] Daniel Grant, "Urban Needs and State Response: Local Government Reorganization" in Alan Campbell, ed., *The States and the Urban Crisis* (Englewood Cliffs, N.J.: Prentice-Hall, 1970), p. 65.

[52] *Ibid.*

[53] Hawley, "Black America's Stake," p. 2.

[54] United States, Advisory Commission on Intergovernmental Relations, *Performance of Urban Functions: Local and Areawide* (Washington, D. C., 1963); Arthur Maass, ed., *Area and Power: A Theory of Local Government* (New York: Free Press, 1959); Werner Hirsch, "Local Versus Areawide Urban Government Services," *National Tax Journal* 17 (December 1964): 331-39.

[55] Robert Dahl, "The City in the Future of Democracy," *American Political Science Review*, December 1967, pp. 955-56, cites literature on economy of scale which leads him to conclude there is no evidence of economies of scale for cities over 50,000 except on a few items such as water and sewage. On methodological difficulties, see Albert Breton, "Scale Effects in Local and Metropolitan Government Expenditures," *Land Economics*, November 1965, pp. 370-72; Harvey Shapiro, "Economies of Scale and Local Government Finance," *Land Economics*, May 1963, pp. 175-81.

do say the money will be spent more efficiently and the costs of not solving areawide problems must also be included in the calculations.[56] Regardless of the criteria used to determine which problems should be handled at regional levels, the resulting lists of areawide functions coincide rather closely.[57] Problems concerned with the physical environment—such as air pollution, water supply, sewage disposal, and transportation—are consistently seen as less local than social problems—such as education and police. Thus the evidence suggests if any functions can be improved by areawide control, they are the physical services; environmental problems are thought to be the most amenable to improvement by metropolitan government.

So, one answer to the question of whether change in structure can improve services for minorities is that the changes most likely to upgrade services will not deal with the services minorities care the most about, namely social services. Of course, minorities have a stake in the improvement of the physical environment, but the issue has less immediate importance for them than jobs and income schemes. Hawley says a multipurpose metropolitan government for the environment would not diminish the power of city officials because they do not have significant control over these functions now.[58] But any strengthened center of power is a potential threat to other power centers simply because it can change priorities. If enviornmental problems of immediate interest to whites receive increased funding, problems of more interest to minorities may be de-emphasized.

Some social scientists maintain social as well as environmental services could be improved by regional government. They point to the necessity of regional solutions to unemployment, housing, education, and segregation problems.[59] For example, metropolitan transportation systems and regional economic development responsive to minority needs could decrease unemployment. Metropolitan zoning and housing programs could alleviate minor-

---

[56]Leslie Carbert, "Financing a Regional Organization in the Bay Area: A Way of Looking at the Problem," Background Paper No. 3, Regional Conference–1970 (Institute of Governmental Studies, University of California, Berkeley, April 18, 1970). Brett Hawkins and Thomas Dye challenge the view that efficiency will result from the elimination of fragmentation. They find fragmentation accounts for little of the variation in public spending, "Metropolitan Fragmentation," *Midwest Review of Public Administration*, February 1970, pp. 17-24.

[57]Bollens and Schmandt, *Metropolis*, p. 166. See the comparison of lists in Stanley Scott and Willis Hawley, "Leadership Views of the Bay Area," *Public Affairs Report*, Bulletin of the Institute of Governmental Studies, University of California, Berkeley, February 1968. For views in the Los Angeles area see David Mars, "Localism and Regionalism in Southern California," *Urban Affairs Quarterly* 2 (June 1967): 47-74.

[58]Hawley, "Black America's Stake," pp. 4-6.

[59]*Ibid.*, pp. 7-12, and Grant, "Urban Needs and State Response," p. 64, and Rendon, "Metropolitanism," p. 8.

ity housing problems. Limited but multipurpose government could lessen fiscal disparities by shifting the financial burden for services such as transportation and recreation to an areawide tax base and thus free some city revenue for services of more immediate concern for minorities. General tax equalization on a regional basis is also advocated as a proper task of regional government which would benefit minorities.[60]

If the evidence on regionalization and the improvement of physical services for minorities is limited, the evidence on social services for minorities is even weaker. Few examples from operating metropolitan governments are given; assertions are made but no demonstrations given. But even if one accepts the view that metropolitan government *can* improve physical and social services for minorities, serious questions arise over whether it actually *will*. First, is a limited multipurpose metropolitan government dealing with social as well as physical problems politically feasible? Second, even if it were formally instituted would minorities actually benefit?

Some social scientists feel metropolitan government with responsibility for social functions is politically feasible.[61] But proposals to strengthen regional handling of physical problems, although opposed by many interests, are infinitely more popular today than proposals to regionalize such functions as housing and zoning, and the equalization of taxes on a regional basis for all metropolitan problems seems extremely unlikely.[62] Oliver Williams emphasizes the resistance of municipalities to integration of services such as land use and education which have high social value to the inhabitants. Thus the trend toward larger governmental units may not portend regionalization of social functions in the next five to ten years. Federal incentives for areawide bodies have stimulated much change, but resistance by localities will undoubtedly become more intense if federal attempts are made to take more social programs away from localities and states and determine how they are handled by metropolitan organizations controlled largely from the national level.

When considerations of feasibility are combined with social science evidence about the effect of other governmental reforms in cities, the possibility that metropolitan government actually will benefit minorities seems extremely remote. Previous reforms in city government, such as nonpartisan elections and council managers, were said to be in the public interest, but a growing body of literature suggests they served the interests of the middle

---

[60]J. M. Banovetz, "Metropolitan Subsidies: An Appraisal," *Public Administration Review* 25 (December 1965): 297–301, says financial gains for central cities have been overrated.

[61]Hawley, "Black America's Stake," p. 15.

[62]*Ibid.*, Even proponents do not claim it is likely, just that it is more likely under metropolitan government than under the status quo.

class. Reformed cities are not responsive to demands from minority groups.[63] Advocates of reform naturally maintain it will serve all people, but if it embodies middle-class values and is implemented by people who accept these values, it is likely to be most beneficial for the middle class. Metropolitan government proposals fit into this category. Some social scientists believe in the ability of an enlightened elite to implement policies which benefit minorities even when they lack resources to organize and influence policies. These analysts see metropolitan government as a neutral instrument which can be put to use to help minorities if the "right" people are in control. They argue that if metropolitan reforms are implemented, minorities will be helped even if they do not have a strong voice. Yet metropolitan government does not seem to be a neutral structure. Larger governmental units necessarily cut down the numerical representation of minorities. For example, at-large elections in American cities typically result in the selection of fewer minority members than district elections. Also, a county that includes many white suburbs is more responsive to those interests than a central city within that county which has large proportions of minority members. It is a mathematical reality that middle-class interests predominate in a jurisdiction where the middle class constitutes the vast bulk of the constituency.[64] So an elected metropolitan government has an inherent tendency to decrease the representation of minorities.

A recurrent theme in political analysis is the ability of those with substantial resources to have the most influence on, and to benefit the most from policy; existing forces tend to get control of new structures.[65] Thus evidence suggests that metropolitan reforms, even if actually instituted, *will not* improve services for minorities (even if in theory they could) unless minorities

[63] Robert Lineberry and Edmund Fowler, "Reformism and Public Policies in American Cities," *American Political Science Review* 61 (September 1967): 701-16; Edgar L. Sherbenou, "Class Participation, and the Council Manager Plan," *Public Administration Review* 21 (Summer 1961): 131-35; Leo Schnore and Robert Alford, "Forms of Government and Socio-Economic Characteristics of Suburbs," *Administrative Science Quarterly* 8 (June 1963): 1-17.

[64] For examples in Los Angeles see David Sears, "Los Angeles Riot Study: Political Attitudes of Los Angeles Negroes" (Los Angeles: Institute of Government and Public Affairs, 1967), pp. 5-6; Dale Marshall, *The Politics of Participation in Poverty* (Los Angeles: University of California Press, 1971).

[65] See David Truman, *The Governmental Process* (New York: Alfred Knopf, 1951) and the community power literature reviewed in Bollens and Schmandt, *Metropolis*, pp. 128-33, such as Robert Presthus, *Man at the Top* (New York: Oxford University Press, 1964) and Peter Bachrach and Morton Baratz, "Two Faces of Power," *American Political Science Review* 56 (December 1962): 947-52. See also Scott Greer, *Urban Renewal and American Cities: The Dilemma of Democratic Intervention* (Indianapolis: Bobbs-Merrill, 1966).

have a strong voice in implementation. Furthermore, even if improved services were possible because of an enlightened elite, we are becoming aware that services cannot compensate for lack of voice, because participation in the process has become essential to minority leaders; they will not cooperate in the services unless they have a voice and with their cooperation improved services are hindered.[66]

To summarize, services to minorities could conceivably be improved by metropolitan government, but the chances are not great unless the minorities have political power. Thus, the feelings and suspicions of minorities described in the previous section can be supported by the evidence. Here we see the essence of the problem of metropolitan government from the minorities' view. Metropolitan government is unlikely to benefit them unless they have a strong voice in it, and yet metropolitan government is a direct threat to the one arena in which they are developing power, namely city government.

Certain advocates of metropolitan reform hope to circumvent the difficulties discussed here by using federal leverage to insist that metro governments benefit minorities. Methods proposed include making federal money for urban social programs contingent upon municipalities giving up powers over land use and education to metropolitan bodies, and requiring minority participation on metropolitan bodies. The federal government is to be the protector of minority interests, using metropolitan government as its vehicle for bypassing states and localities. Doubts were expressed earlier about how successful such a federal strategy can be in the face of intense state and local opposition. Also, the motivation of the federal government to fulfill this role will undoubtedly vary from one administration to another and will reflect changes in the national political climate. Thus minorities cannot rely upon federal action to overcome decreasing minority power in cities. Many groups—state, local, and national—are trying to control the governance of regions. The outcome is not certain, but it will determine how the tensions between the power of minorities and the quality of the services they receive are resolved.

Conclusion

Given this interpretation of the conflict between service and power for minorities in metropolitan government, the temptation is to give in to despair, or to switch to some other, more consoling vocation. However, assuming that the gradual emergence of larger units of government is going to continue, a more promising response is to focus on strategies for minorities. Taking as given the tension between minority power in cities and the trend toward metropolitan

---

[66] Daniel Bell and Virginia Held, "The Community Revolution," *The Public Interest* 16 (Summer 1969): 142-79.

government, there may be ways of minimizing the negative results for minorities.

One strategy is to criticize metropolitan government, ignoring or opposing all local, state, and national efforts to implement it, and at the same time to demand community control. The purpose is to consolidate power in cities or subareas of cities before getting involved in other areas. There are several arguments against this approach.[67] Since black control of most cities is not an immediate prospect, the wait before involvement in the metropolitan arena will be long. In the meantime the development of metropolitan government is already under way, and minorities are underrepresented in this process. For example, of 170 positions on 11 metropolitan districts in the San Francisco area, only 3 were filled by minority members.[68] And finally, the significance of political power in the cities or subareas of the cities is declining. So even in the handful of cities in which blacks can expect to have control in the next fifteen years, the victory may be hollow. Short-term success at getting power in cities may not enable minority leaders to improve the condition of their constituents' lives. Cities have limited powers over many issues, such as employment, which significantly affect the quality of life for minorities. And the cities are dependent on state and national financing for many functions such as schools, housing, and welfare.[69]

In the light of these considerations, a more promising strategy is to continue criticizing metropolitan government, but to stop ignoring it. Minorities could become active in the efforts to shape metropolitan reorganization at local, state, and national levels—opposing proposals detrimental to minorities and devising alternative proposals in order to be included in the bargaining process. Much attention needs to be given to the content of the alternative plans; and they will vary according to the situation in each metropolis and each state, depending on the existing power of minority groups and the possibilities for coalition. Since whites have become more interested in regional solutions to environmental problems, one promising approach might be for minorities to withhold support for such solutions until they are assured that the multipurpose government will handle certain aspects of the social problems that are not amenable to mere local control, yet permit community control over the other aspects. For example, minorities could demand a multipurpose district which equalizes the tax base for education yet places the control of schools at the community or city level (depending on the

---

[67] Hawley, "Black America's Stake," pp. 17-20.

[68] Brown, "Regional Government."

[69] Paul Friesma, "Black Control of Central Cities: The Hollow Prize," *Journal of American Institute of Planners* 35 (March 1969): 76-79; Roland Warren, "Politics and the Ghetto System," in *Politics and the Ghetto* (New York: Atherton Press, 1960), p. 25; Arnold Schucter, *White Power/Black Freedom* (Boston: Beacon Press, 1968), p. 374; and Rendon, *Metropolitanism*, p. 13.

degree of power the minorities have at each level). Altshuler convincingly argues the desirability of such a combination of regional action with community control and the possibility of forming a coalition between advocates of centralization and community control.[70] Since coalitions with middle-income interests are essential, the best trade-offs must be determined. For example, minorities may decide to support pollution programs in return for white support of transportation systems designed to serve ghetto interests. In addition to making proposals about the function of metropolitan government, minorities should specify the type of representational system required in order to gain minority support for metropolitan reorganization. Direct district elections with districts purposely designed to insure the election of minority representatives would be most likely to give minorities power and make the new government responsive to minorities after it is created.[71]

The proposal here for minimizing the negative effects of the tension between minority power in cities and the trend toward metropolitan government can be called participatory opposition. It involves using the existing city power of minorities in both local and national politics as leverage to decrease the threat that metropolitan government could pose to that power—in other words, mobilizing to shape a metropolitan government which will provide the power minorities need to influence the quality of the services provided by that reorganized government. Piven and Cloward argue that the Negro community cannot be organized to bargain effectively.[72] But Aberbach and Walker present strong evidence of the increased ability of Negroes to organize; they describe a "more unified, more highly mobilized black political community."[73] Accepting the ability of Negroes and other minorities to bargain effectively does not justify the naive conclusion that all their demands will be met. As discussed earlier, the political feasibility of some of these demands are in doubt because they will come into conflict with demands from other groups, most obviously the white suburbs. But participation by minorities in the bargaining will improve the chances that the forms of metropolitan government that emerge will have been influenced by the

---

[70] Altshuler, *Community Control*, pp. 47–50; and Bell and Held, "The Community Revolution," p. 176; Piven and Cloward, "Black Control," October 7, 1967, p. 18. For additional comments relevant to relations between larger units of government and smaller units, see Robert Dahl, *After the Revolution* (New Haven: Yale University Press, 1970), pp. 140–65; James Sundquist, *Making Federalism Work* (Washington, D.C.: The Brookings Institution, 1969), pp. 240–41; and Theodore Lowi, *The End of Liberalism* (New York: Norton, 1969), pp. 194–99, 305–6.

[71] The argument here follows Hawley, "Black America's Stake," p. 21.

[72] Piven and Cloward, "Black Control," October 7, 1967, p. 18.

[73] Joel Aberbach and Jack Walker, "The Meanings of Black Power: A Comparison of White and Black Interpretations of a Political Slogan," *American Political Science Review* 64 (June 1970): 387.

desires of minorities. The argument is that the compromises necessitated by participatory opposition are preferable to the results likely from a strategy of minority isolation.[74]

The present review of the attitudes of minorities to metropolitan government and of the possible effects of metropolitan government on minorities leads to three main conclusions. First, those interested in any form of metropolitan government and in the plight of minorities ought to pay more attention to the negative reactions of minorities. They serve as important reminders of lessons that reformers should have learned from the results of past efforts. Structural reforms are not neutral. Depending on the values they reflect and the groups that get control of them, they benefit certain interests and hurt others. Metropolitan government in many of its most popular forms could dilute the voices of minorities, as other city reforms have in the past. Further, structure alone does not determine function. The mirage of reorganization must not blind us to the way existing forces tend to get control of new structures. The voices of other, new forces must be purposely strengthened if the innovative potential of the structure is to be realized. Those interested in improving the position of minorities can consistently favor metropolitan government as well, if they are willing to design the reform to counteract the possibility that minorities will be harmed.

Second, active participation in the evolution of metropolitan government by minorities would help make the reformers take account of these neglected interests. A review of the literature of metropolitan reform illustrates beautifully the way scholarly analyses reflect the intellectual climate of the times. Since much of America's history has been characterized by a disregard for the problems of minorities, it is not surprising that discussions of metropolitan reform showed a similar tendency until the whole society recently "became aware" of race.[75] But just as metropolitan reformers have often been oblivious to minorities, minorities have paid little attention to metropolitan government. In the outpouring of literature written by minorities about minority problems, very few mentions are made of trends toward larger units of government.[76] Minorities need to study the potentially threatening trend and to

---

[74] And some of the compromises may not be harmful to minorities. For instance, Victor Jones says a mixed system of representation combining direct election with constituent unit representation will actually increase the representation of minorities in the San Francisco area. "Representational Local Government: From Neighborhood to Region," *Public Affairs Report*, Bulletin of the Institute of Governmental Studies, University of California, Berkeley, April 1970.

[75] Walter Stafford and Joyce Ladner make this same point about the planning field in "Comprehensive Planning and Racism," *Journal of the American Institute of Planners* 35 (March 1969): 68-74.

[76] See *The Black Politician, The Civil Rights Digest*, and *Mexican American Journals* cited on pp. 96-100 of the *Guide to Materials Relating to Persons of Mexican Heritage*

consider ways of countering the threat. The views of black social scientists are needed to insure that diverse perspectives are gained.

Third, and finally, the incomplete and impressionistic evidence on the relation of metropolitan government to minorities makes the need for more study painfully obvious. We need to look systematically at the attitudes of many minorities in different cities, and of many segments of each minority community, toward proposed and existing radical and incremental reforms. Are the San Francisco findings of minority desire for multipurpose government dealing with social problems and elected by districts duplicated in other cities and with larger samples? What are the effects of existing radical and incremental reforms in diverse metropolitan centers on minorities' services and power? For example, does widening a given tax base help the minorities in the central cities? How does the operation and output of single-purpose districts compare with those of multipurpose districts in regard to minorities?[77] What type of metropolitan functions and structures are most responsive to minorities? For example, what forms of metropolitan government would help solve each of the problems minorities face, problems such as housing, jobs, education, welfare, and so on? What kinds of power over which functions are most promising for community control, and how can it be combined most effectively with metropolitan government? What strategies in what settings are most successful in gaining minorities' power on regional government? For example, how should strategies vary in cities where blacks are in control and in cities where they will be in the minority at least for the next fifteen years?

The list of questions for study could be elaborated endlessly. The general topic of metropolitan government is old, but as our political system evolves, the topic takes on new meanings, and we have not kept up with these changes. In the midst of a society of ever-increasing scale, metropolitan government is no longer a reformer's proposal to be approved or defeated at the polls—its existence in our federal system is becoming increasingly apparent and important. The question is not whether there will be larger units of government, but (1) what kinds will they be? and (2) what impact will they have on our domestic problems, especially the position of minorities? Forces are jockeying now to see how these questions will be answered. The importance of the issues involved underscores the need for new interest among social scientists and minorities in the problems of metropolitan government's impact on minorities. Their concern could contribute to the resolution of these problems.

in the United States, compiled by the U.S. Inter-Agency Committee on Mexican American Affairs (Washington, D.C.: Government Printing Office, March 1969).

[77]Warren, *Politics and the Ghetto*, p. 7, comments on the lack of data on the political forces affecting ghetto policies.

2 Blacks in Suburbia:
The Myth of Better Opportunities

BERNARD J. FRIEDEN*

The warning issued by the Kerner Commission in 1968–"Our nation is moving toward two societies, one black, one white–separate and unequal"– has its territorial counterpart in the racial divisions within our urban areas. With blacks concentrated in the central cities, surrounded by overwhelmingly white suburbs, the turf held by each of the two societies seems all too clear. That these two territories offer their respective residents a very unequal set of resources also seems clear. As a result, many students of urban affairs have argued that where black people live is not merely an accidental outcome of the way the housing market operates, but has become in itself a way of reinforcing the disadvantages blacks face in American society. If this view is accurate, opening the suburbs to the black poor should be a prime aim of public policy, not only because freedom of movement is a fundamental right in itself, but also because it is an important way of reducing racial inequalities. Presumably the key to a suburban house will also unlock a wide range of opportunities that are now closed to black people.

The political behavior of white suburbanites suggests that they may share this view. Even liberal communities whose residents are willing to send their

*Professor, Urban Studies and Planning, Harvard University, and Director, Joint Center for Urban Studies of the Massachusetts Institute of Technology and Harvard University.

Note: For support of preliminary research for this paper, I am indebted to the Urban Ghetto Study Program sponsored by the U.S. Economic Development Administration at the Laboratory for Environmental Studies at the Massachusetts Institute of Technology, and to the M.I.T.-Harvard Joint Center for Urban Studies. Elaine R. Savitsky was most helpful as a research assistant.

tax dollars into ghetto neighborhoods are often unwilling to share their living space or their classrooms with blacks from the central cities. Local opposition to open occupancy and low-income housing has prompted some observers to conclude simply that liberalism stops at the driveway. It is also possible that many suburban voters believe that the present territorial division of metropolitan areas does indeed protect their own social and economic status, and will give their children a head start in the race for the good things in life.

In the next decade, the opportunities for black people to move to suburbia will be better than they have been for a long time. During the 1970's, an increasing stock of older suburban housing is likely to be available at prices that many black families can afford. Much of this housing dates from the years immediately following World War II, when the volume of suburban homebuilding reached new peaks. (The best year the homebuilding industry ever had was 1950.) These houses will be 20 to 25 years old, built on small lots and with interior space well below what middle-income home buyers now expect. Also, many of them will show some effects of wear and tear. Thus, if new construction picks up once again to take care of the demand from middle- and upper-income groups, and if mortgage financing is available at reasonable rates, these older homes could make it possible for a large number of families with modest incomes, black or white, to move to the suburbs. At the same time, the volume of federally subsidized low-and moderate-income housing is likely to be several times greater than it was in the 1960's. By 1970, the total number of new subsidized units was more than five times greater than it had been in the mid-1960's, with further increases promised by the Nixon Administration. If, as seems likely, a substantial share is built in the suburbs, this will serve as another important resource for low-income families who want to locate in suburbia.

Although the opportunities for black people to live in suburbia may be expanding, some groups that want to promote racial equality have begun to question whether opening up the suburbs is a fruitful objective for public policy. In particular, the political ideology created by black militants argues that movement to the suburbs at this time will not work to the advantage of black people as a whole. Although the militants are certainly not creating obstacles to such mobility (white people can be counted on to continue in this role), they contend that first priority should be given to establishing greater self-government in the central-city black communities and greater group solidarity. Further, they maintain that a wider dispersal of blacks to the suburbs will weaken their voting bloc in the central cities.

The possibilities for change in the racial pattern of urban areas, as well as the fresh doubts about what such change would achieve, make it timely to take a closer look at this subject. The basic issue is whether an accelerated movement of black people to suburbia would reduce the disadvantages they now face in the central cities and, ultimately, whether such a movement would bring about greater social and economic equality. More specifically, the question is whether, by achieving a residential distribution similar to that of white people, blacks will be able to get similar access to good housing, public services, and employment. If this were to happen, we would not necessarily eliminate poverty, slum housing, or service disparities, but these conditions would be no more prevalent among blacks than among whites. In that event, we would still have problems of social inequality, but they would no longer be compounded explosively by problems of racial inequality.

A closer look at the possible impact of suburban residence upon black people will also make possible more considered judgments about priorities for public policy. One of the issues that has so far received little attention is the possible need for public action beyond opening up the suburbs; that is, action to help black families after they are already living in the suburbs. The various arguments that have been advanced in favor of suburbanization as a solution to the problems of black people can be tested against available data. Most of the data are fragmentary and suffer greatly from definitional difficulties. Nevertheless, they do suggest that the case for suburbanization needs to be qualified in several important respects, and that a different residential pattern in itself will make only limited contributions toward closing the gap between the races.

In raising questions about the possible social and economic effects of greater suburbanization among the black population, I do not mean to imply that public policy should be based solely on these issues. Far from it: even if it should be found that blacks fare very poorly in suburbia, they still have an obvious right to live there or anywhere else. No matter what the social and economic consequences may turn out to be, public policy must work toward establishing and safeguarding freedom of residential choice for all people. The results of this investigation cannot in any way argue against a governmental commitment to opening the suburbs to black and other minorities as a basic matter of civil rights. The only acceptable basis for comparing the benefits of central city and suburban residence must be that living in either place is to be a matter of individual choice rather than restricted opportunity.

Movement of Blacks from City to Suburb

Current trends in population movement provide the context in which public policies to help blacks enter the suburbs would have to operate. The black population of urban areas continues to expand rapidly, primarily through natural increase rather than migration. This increase (currently estimated as 347,000 per year) supplies the main push to find additional living space.

Space can be found either in established black neighborhoods, other parts of the central cities, or in the suburbs. Many of the older black areas, however—including Harlem, Hough, and Watts—have been losing population for some time. Most of the population increase has been accommodated in other neighborhoods of the central cities, while whites have been withdrawing from these areas to suburbia, giving rise to the distinct possibility that we are moving toward two societies, geographically.

In the early 1960's, the growth rate of black population in the suburbs fell to a very low level, even below the annual rate of change for the 1950's. Census Bureau estimates for 1960-64 were that the growth of Negro suburban population failed even to keep pace with natural increase, and that there was probably a net migration of blacks out of the suburbs during this period. Had this pattern continued, we would clearly be fulfilling the prophecy of the Kerner Commission. But by the mid-1960's this trend began to change. The first signs of change, coming from the Current Population Survey of 35,000 households throughout the nation, were discounted as possibly erroneous. But repeated surveys, an enlarged national sample, and supporting evidence from other sources now indicate more firmly that a significant growth of black population is under way in the suburbs.

The most recent Census Bureau estimate is that Negro population in the suburbs of metropolitan areas has been growing by an average increment of 85,000 per year from 1964 to 1969, of which 34,000 represents net inmigration from other places.[1] For the period 1960 to 1969, the same estimate indicates that Negro population increased at a slightly faster rate in the suburbs than in the central cities (33 percent to 27.7 percent). If these figures are accurate, black people are not concentrating increasingly in the central cities: they show a slightly higher proportion of the total Negro population of metropolitan areas in the suburbs in 1969 than in 1960.

Some corroboration of decreasing racial separation comes from data gathered by the University of Michigan's Survey Research Center in 1964 and 1968. In national surveys, the percentage of whites who said they live in all-white neighborhoods dropped from 80 to 75 percent, and the proportion of blacks who reported living in all-black neighborhoods went from 33 to 25 percent. In parallel results, the proportion of whites who reported that the high school nearest them was all-white declined, as did the proportion of blacks reporting their nearest high school as all-black.[2] These data do not, however, report directly on movement of blacks to the suburbs.

---

[1] U.S. Bureau of the Census, *Current Population Reports*, Series P-20, No. 197, 1970, p. 3. See also David L. Birch, *The Economic Future of City and Suburb* (New York: Committee for Economic Development, 1970), pp. 28-33.

[2] University of Michigan, Institute for Social Research, *ISR Newsletter* 1 (Winter 1970): 4.

A very gradual shift of black population from central cities to suburbs should not be surprising in view of earlier trends. On the basis of 1950-60 experience, projections prepared for the Douglas Commission anticipated that by 1985 there would be a small increase in the proportion of metropolitan nonwhite population located in the suburbs. The Douglas Commission projections showed an average annual increment of 160,000 nonwhites in the suburbs over this period.[3]

A comparison between blacks and whites in the same income brackets would also suggest the likelihood of accelerated black movement to the suburbs. For some time, the white poor have been making their way into suburbia. By 1960, 44 percent of white families in metropolitan areas with incomes below $3,000 lived outside the central cities (compared with only 14 percent of nonwhite families with similarly low incomes). One study made use of 1960 Census data to show what would happen if, at every income level, blacks were to become homeowners in the same proportion as whites and if, at every income level, blacks were to distribute themselves between central cities and other parts of metropolitan areas in the same proportions as whites. In eleven of the largest metropolitan areas for which these calculations were made, the proportions of nonwhites living in the suburbs would go from an actual 16 percent as of 1960 to 40 percent.[4] These calculations demonstrate that at the income levels that blacks had reached by 1960 (which are still higher now), whites were finding a plentiful supply of housing in the suburbs.

Nor is there much evidence to suggest that blacks themselves have strong preferences to remain in predominantly black neighborhoods in the central cities. Aside from the fact that more blacks have been "voting with their feet" to go to suburbia recently, opinion polls continue to show that only a small number favor living in all-black areas. A survey conducted for the Kerner Commission in fifteen cities asked, "Would you personally prefer to live in a neighborhood with all Negroes, mostly Negroes, mostly whites, or a neighborhood that is mixed half and half?" Of the Negroes interviewed, 48 percent said mixed half and half, 37 percent said it makes no difference, and only 13 percent preferred all or mostly Negro neighborhoods.

In summary, then, the push of black population growth in metropolitan areas continues to create pressures for living space beyond the traditional ghetto neighborhoods. The income levels of black families match the cost of housing in the suburbs well enough to permit substantial movement there, and there are signs that this movement is now under way. Public policies

[3] Patricia Leavy Hodge and Philip M. Hauser, *The Challenge of America's Metropolitan Population Outlook, 1960-1985*, U.S. National Commission on Urban Problems, Research Report No. 3 (Washington, D.C., 1968), p. 26.

[4] Richard Langendorf, "Residential Desegregation Potential," *Journal of the American Institute of Planners* 35 (March 1969): 90-95.

concerned with promoting greater racial equality could conceivably ignore such movement and concentrate on other matters, or it could attempt to accelerate this trend. In addition, public policy could recognize suburban blacks as a growing constituency that may need special help, or it could focus primarily on the much larger group of blacks who will continue to live in central cities.

The Suburbs and Access to Jobs

One of the most important arguments advanced in favor of promoting the movement of blacks to suburbia is that low-income people will have much better access to jobs if they live there than if they live in the central cities. The number of jobs has been growing much faster in the suburbs than in the central cities, particularly in manufacturing and retail trade. Transportation is poor between black neighborhoods in the central cities and the new suburban job locations, so that workers either have to drive long distances or make cumbersome and expensive trips by public transit. Further, several studies have shown that industrial workers find jobs primarily through informal channels of communication—through knowing others who work at the same plant or through living nearby and hearing of job vacancies. Thus the spatial separation between central-city blacks and suburban jobs, coupled with their lack of contact with other suburban workers, may prevent them from even learning about openings for which they are qualified.

John Kain's analysis of the relationship between housing segregation and employment in Detroit and Chicago lends some empirical weight to this argument. Using data from 1952 and 1956, he found Negroes systematically underrepresented in work places far from where they lived. Kain estimates that Negroes could have held as many as 25,000 additional jobs in Chicago and 9,000 additional jobs in Detroit if their residences were dispersed more widely through these metropolitan areas.[5]

Nevertheless, the various components of this argument do not point unequivocally to the conclusion that movement of blacks to suburbia will go a long way toward improving their jobs and raising their incomes. Although it is true that jobs have been growing much faster in the suburbs, the total number of jobs remains greater in the central cities. Blacks living in these cities could conceivably improve their economic position by going after the large volume of job openings that result from normal labor turnover, especially since the competition from whites may be lessening as the departing white labor force becomes more oriented to the suburban job market. And in most sectors of the economy, the decline of the central city is relative, not

[5] John F. Kain, "Housing Segregation, Negro Employment, and Metropolitan Decentralization," *Quarterly Journal of Economics* 82 (May 1968): 175-97.

absolute. Between 1958 and 1967, for example, retail sales increased at a real rate of 13 percent in the central cities of the thirty-seven largest metropolitan areas, while suburban retail sales increased by 105 percent.[6] Further, most data series on jobs in central cities and suburbs fail to include government employment. In the thirty largest metropolitan areas total local government employment increased enough between 1957 and 1962 to more than offset job losses in other sectors. After 1962, the decline in private employment was reversed in the central cities of most of the large metropolitan areas: in the thirty largest areas, taken together, private employment rose by an average of 99,000 jobs per year between 1963 and 1967, compared with the average annual decline of 6,000 between 1958 and 1962.[7] Conceivably, a job training and placement strategy to open more central-city jobs to blacks, particularly in the growth sectors of office and public service employment, could be as effective in reducing racial disparities as a policy of helping blacks move to the suburbs.

Nor is it entirely clear that black people in the suburbs are able to improve their job status more easily than black people in the central cities. About the same proportion of Negro families were below the poverty level in both central cities (23 percent) and suburbs (24 percent) in 1968.[8] Unemployment rates for blacks in the suburbs are not consistently or significantly lower than those for central-city blacks. One recent survey indicated slightly higher rates of unemployment for Negroes living outside the central cities than for those within: for men, 7.1 percent to 6.9 percent; for women, 8.6 percent to 8.2 percent.[9] A survey of unemployment in the twenty largest metropolitan areas indicated a rate of 6.3 percent for nonwhites in the central cities, and 5.3 percent in the suburbs.[10] The differences between central cities and suburbs are not very great in either case, and the data certainly do not show a clear advantage for suburban residents in this respect. This finding does not contradict in any way the fact of very high unemployment rates in many of the ghetto areas of central cities, or the still higher rates of unemployment in these areas revealed by the Labor Department's index of subemployment

[6]U.S. Advisory Commission on Intergovernmental Relations, Bulletin 70-1; "Metropolitan Disparities—A Second Reading" (Washington, D.C., January 1970), p. 3.

[7]Benjamin I. Cohen and Roger G. Noll, "Employment Trends in Central Cities" (unpublished manuscript), p. 20. The metropolitan areas are Baltimore, Denver, New Orleans, Philadelphia, San Francisco, St. Louis, and Washington.

[8]U.S. Bureau of the Census, *Current Population Reports*, Series P-23, No. 29 (1970), p. 23.

[9]U.S. Bureau of the Census, *Current Population Reports*, Series P-20, No. 175 (1968), p. 6.

[10]U.S. Bureau of the Census, *Current Population Reports*, Series P-23, No. 29 (1970), p. 39.

(which counts people not usually defined as being the labor force, plus those working part time, and those working full time but earning incomes below the poverty level). But the striking rates of ghetto unemployment seem to be offset by much lower rates among blacks living in other parts of the central cities—and these other neighborhoods must be seen as the logical alternative to the suburbs as places to accommodate growing black populations.

Little information is available on income differentials for blacks in central cities and suburbs. One set of data on this subject indicates that 1968 median incomes for Negro males were higher in the suburbs than in the central cities for metropolitan areas of 1 million or more, but were lower in the suburbs than in the central cities for metropolitan areas of less than 1 million. Median incomes for Negro females were about the same in all parts of the large metropolitan areas, but were lower in the suburbs than in the central cities of the smaller metropolitan areas.[11] As of 1966, median incomes for blacks were higher in the suburbs than in the central cities ($5,535 to $5,171), taking all metropolitan areas together, but the ratio of black to white median incomes was lower in the suburbs (0.62 to 0.66).[12]

In terms of occupational classifications, the 1960 Census showed an uneven pattern in different parts of the country. The proportion of professional and technical employees among nonwhites was higher in suburbs than in central cities in the Northeast (9 percent to 5 percent) and the North Central states (9 percent to 4 percent), but the same in the West (8 percent in both) and lower in the suburbs in the South (4 percent to 5 percent).[13]

These data, fragmentary as they are, show little support for the belief that suburban residence is the key to better jobs and higher incomes. They suggest instead that residential location in itself is only one factor of many that contribute to racial inequalities in jobs and incomes, and that other factors may be more important singly or certainly in combination. Conceivably, even living in the suburbs, closer to the centers of new job growth, does not suffice to bring black people into good communication with the job market. If public transportation between central-city locations and the new industrial parks is poor, it is probably no better in the suburbs. The worker without a dependable car is probably equally disadvantaged in both places. Or perhaps living in suburbia does help people cope with problems of communication or transportation, but it does not provide workers with new job skills nor does it

[11]U.S. Bureau of the Census, *Current Population Reports*, Series P-60, No. 66 (1969), pp. 85, 87.

[12]U.S. Bureau of the Census, *Current Population Reports*, Series P-20, no. 175 (1968), p. 2.

[13]U.S. Advisory Commission on Intergovernmental Relations, *Metropolitan Social and Economic Disparities* (Washington, D.C., January 1965), p. 172.

deal with discrimination in hiring or in the selection of people for manpower training programs. Opening more suburban housing to the black poor may be helpful in terms of employment, but the evidence suggests that it is not decisive.

Access to Schools and Public Services

The case for aggressive public action to open the suburbs to larger numbers of blacks rests also on the argument that living in the suburbs will give people access to superior schools and higher quality public services than they can get in the central cities. Judging the quality of schools and other services is difficult, but the level of expenditures on these items can offer some insight into the validity of this argument.

When the suburbs of metropolitan areas are grouped together, it is clear that they consistently spend more per pupil on education than the central cities. In the thirty-seven largest metropolitan areas, the suburbs spent more per pupil in 1964–65 than the central cities in all but three cases (in two of these the expenditures were the same for central cities and suburbs, and in only one case, Denver, did the central city spend more).[14] Average expenditures per pupil in the thirty-seven areas were $449 in the central cities and $573 in the suburbs. Moreover, the central cities have been falling further behind since then.

It is important to note, however, that there are wide variations from one suburban school district to another, and that many suburbs spend less per pupil than the central cities. In metropolitan Los Angeles, for example, central city expenditures of $576 per pupil in 1965–66 were far exceeded in the suburbs of Pasadena ($698) and Beverly Hills ($975); but Pomona spent only $541, Baldwin Park, $522, West Covina, $502, and the lowest ranking school district in the County spent $478.[15] The argument must be qualified at least to the extent of specifying that a selective movement of blacks into suburbs whose school outlays are close to the average will give them access to better schools than those of the central city, on the assumption that money can buy good education.

Whether suburban schools are more effective than central city schools in meeting the needs of students who have educational problems or who want vocational education rather than college preparation is less clear. In 1960, the percentage of high school dropouts (16- and 17-year-olds not enrolled in school) was higher in the central cities in the Northeast and North Central

---

[14]U.S. Advisory Commission on Intergovernmental Relations, *Fiscal Balance in the American Federal System*, Vol. 2, *Metropolitan Fiscal Disparities* (Washington, D.C., October 1967), p. 66.

[15]*Ibid.*, p. 346.

regions, but higher in the suburbs of the South and West.[16] One major study of central city-suburban disparities found that suburban school districts generally lack appropriate facilities or staff for adequate vocational education, whereas the large central cities were better equipped in this respect.[17]

Educational levels among black adults who now live in the suburbs may reflect opportunities in the suburbs, or they may merely reflect opportunities in other communities where they lived when they were of school age. In any event, the median school years completed by Negroes aged 25 and over in 1969 were higher in the central cities (10.6) nationally than in the suburbs (10.2). Of those living in the central cities, 61.9 percent had completed less than 4 years of high school, while 63.8 percent in the suburbs had less than a high school education.[18]

Suburban schools can offer an educational setting that may be significant for black students. The Coleman Report (*Equality of Educational Opportunity*)[19] and subsequent analyses of its data by the U.S. Commission on Civil Rights found the class and racial composition of schools to be the most significant factors correlated with the performance of Negro students. The integration of black students into schools with white majorities seemed to make the greatest positive impact on their achievement. This study has given rise to a great debate in educational circles, and it has been criticized on methodological and other grounds. If its findings are valid, however, they would argue for educational policies that can best be carried out in the suburbs. In many of the largest metropolitan areas, blacks constitute so high a proportion of the central-city school population that it has become virtually impossible to organize integrated schools in the sense of schools with white majorities. The techniques of bringing about school integration simply will not work when there are not enough white students to go around. In the suburbs, however, this is rarely the case. We do not have good data on the distribution of black population in the suburbs, but there appear to be very few suburban communities in which black students predominate. Although these students frequently live in segregated neighborhoods within the suburbs, such devices as redistricting, busing, and the use of large educational parks are capable of bringing about school integration when black students are a minority. When black families are segregated within the suburbs, how-

[16] U.S. Advisory Commission on Intergovernmental Relations, *Metropolitan Social and Economic Disparities*, p. 166.

[17] *Ibid.*, p. 117.

[18] U.S. Bureau of the Census, *Current Population Reports*, Series P-20, No. 194 (1970), p. 16.

[19] James Coleman, *Equality of Educational Opportunity* (Washington, D. C.: Government Printing Office, 1966).

ever, the potential advantages of integrated education will not be realized unless there is deliberate public action to integrate the schools.

If expenditures on schools indicate that blacks can get a better education in suburbia, then expenditures on other public services suggest that their other needs may be met better in the central cities. Noneducational expenditures of local governments, on a per capita basis, are consistently higher in the central cities. In the thirty-seven largest metropolitan areas as of 1966-67 there were no exceptions to this pattern; the average amounts were $230 in the central cities and $138 in the suburbs.[20] Further, the gap between central city and suburban spending on noneducational functions has been growing since the 1950's.

Comparative information on specific public services is scarce. Data from the largest metropolitan areas, however, show consistently and substantially higher central city spending for welfare, police, fire protection, health and hospitals, and for water, sewer systems, and other utilities.[21] Limited data on the distribution of federal aid also suggest that the central cities are apt to provide more elaborate programs relevant to the needs of poor people. Information is available on federal aid to eight large metropolitan areas in 1967. The central cities, with about 50 percent of total population, received 57 percent of federal funds in all, and 67 percent of federal aid other than FHA mortgage insurance. In programs linked directly to poverty, the central cities received 81 percent of the funds distributed by the Office of Economic Opportunity and 83 percent of public housing funds. They also received 72 percent of Department of Health, Education, and Welfare allocations, 76 percent of urban renewal aid, and 66 percent of Department of Labor funds.[22]

The black family moving from an average central city to an average suburb, then, is likely to find a different mix of public services which is not uniformly better. For a family with steady income and enough resources to have no need of health and welfare aid or antipoverty programs, little will be

[20]U.S. Advisory Commission on Intergovernmental Relations, Bulletin 70-1, "Metropolitan Disparities," Table VIII.

[21]See U.S. Advisory Commission on Intergovernmental Relations, *Fiscal Balance in the American Federal System*, Vol. 2, *Metropolitan Fiscal Disparities*, pp. 71-73, 106-9; and U.S. Bureau of the Census, *Local Government Finances in Selected Metropolitan Areas in 1965-66* (1967), for the following metropolitan areas in which the central city is a separate county with data separately available: Baltimore, Denver, New Orleans, New York, Philadelphia, San Francisco, St. Louis, and Washington.

[22]U.S. Office of Economic Opportunity, *Summary Federal Programs FY 1967* (Washington, D.C.: Government Printing Office, 1968); for the eight metropolitan areas cited in note 21. Data are tabulated in Cohen and Noll, "Employment Trends in Central Cities."

lost, and the school system is likely to be better. For a family coping with low income and the problems that tend to accompany it—periodic unemployment, poor health and no medical insurance, and job skills that need upgrading—the middle-income suburb may be a source of little help. If, in addition, the schools turn out to be unprepared to deal with the special problems of lower-class youth, the black poor may find that living in suburbia gives them access to services that are of high quality for others but irrelevant to them. Conceivably, the suburbs might reorganize their public services to match the needs of an influx of poor black families; but resistance can be anticipated from suburban voters who want to avoid tax increases, particularly for poverty-linked services that could attract additional low-income residents.

Housing and Neighborhood Choices

One of the fundamental beliefs underlying the case for public action to open the suburbs is that black people can improve their housing conditions substantially by moving there. A related view is that black families who want to live in socially and racially integrated neighborhoods will find it much easier to do so in the suburbs than in the central cities.

As for the question of housing quality, there is no doubt that suburban houses are, on the average, newer and in better condition than those in the central cities. In addition, higher proportions of housing units are single-family and owner-occupied in the suburbs. Whether black families who move to the suburbs will find their way to desirable housing is less clear, however.

In the country as a whole, higher proportions of nonwhites live in substandard housing in the suburbs (16 percent) than in the central cities (9 percent), according to Census Bureau estimates for 1968.[23] This pattern results in part from definitional problems: particularly in the South, the "suburban" parts of metropolitan areas include a great many rural and semirural communities in the outlying sections of metropolitan counties, where housing conditions are very poor. The differential between suburban and central city housing is indeed greatest in the metropolitan areas of the South, where 22 percent of nonwhite families and only 9 percent of central city nonwhite families live in substandard housing. But the gap is significant in the North and West, as well: 12 percent in the suburbs and 9 percent in the central cities.

In the North and West, the most plausible explanation is somewhat different. First, metropolitan areas tend to include a number of old industrial communities outside the central city which have many of the same character-

[23] U.S. Bureau of the Census, *Current Population Reports*, Series P-23, No. 29 (1970), pp. 57–58.

istics as the central cities themselves. Census data for 1960 indicated that a substantial proportion of the suburban black population was concentrated in communities with populations of 50,000 or more, where the labor force was engaged mainly in manufacturing. As of 1960, twenty-one communities of this kind contained 300,000 nonwhites of a total 1.3 million nonwhites living in all suburbs of the North and West. Further, even in communities that are more like the stereotypical residential suburb, there are often wide variations in housing quality. In metropolitan Los Angeles, for example, Pasadena is one of the few suburbs that has had a substantial influx of black families. As of 1960, slightly more than 9 percent of the housing units in Pasadena were classified as "deteriorating" or "dilapidated," compared with slightly less than 9 percent in these same categories in Los Angeles itself.[24] Certainly Pasadena is a more promising place to look for good housing than in ghetto areas such as Watts, but it is no more promising than the many outlying neighborhoods of Los Angeles.

In most of the large metropolitan areas, departing white residents have been leaving behind a large supply of housing in reasonably good condition, and this housing has served as the main resource for expanding black populations during the past twenty years. It continues to constitute the main alternative to suburbia for black families, and it has helped many to improve their housing conditions. Since 1960 there has been a substantial decrease in the number of black families in metropolitan areas living in substandard housing, of which about 80 percent has occurred in the central cities.[25]

The suburbs do seem to offer greater opportunities for homeownership than central-city neighborhoods. In 1960, a considerably higher proportion of nonwhites lived in owner-occupied housing in the suburbs (52 percent) than in the central cities (31 percent).[26] Ownership is an important housing option, particularly for families with modest incomes, for several reasons. It is a way of escaping the frustrations of a difficult bargaining position with landlords over service, rent, and even continued occupancy—all of which have been especially troublesome recently in ghetto neighborhoods. For the low-income family, ownership can also serve as a form of savings and an investment. In this respect, an accelerated movement of black families out of the central cities may be a way of promoting greater racial equality.

When families search for a house, many are as concerned with who lives in the neighborhood as they are with the quality of the house itself. Black

[24] U.S. Advisory Commission on Intergovernmental Relations, *Fiscal Balance in the American Federal System*, Vol. 2 *Metropolitan Fiscal Disparities*, pp. 352, 354.

[25] U.S. Bureau of the Census, *Current Population Reports*, Series P-23, No. 29 (1970), p. 60.

[26] U.S. Advisory Commission on Intergovernmental Relations, *Metropolitan Social and Economic Disparities*, p. 182.

families presumably will be no exception. A sizable number are likely to want to live in neighborhoods where working-class or middle-class life styles predominate, in contract to the lower-class life styles of inner-city poverty areas. Some will undoubtedly prefer to live in neighborhoods that are racially mixed as well. The suburbs seem to offer better prospects for finding such areas, although the evidence suggests that the suburban advantage is not always great.

Families with incomes below the poverty level do appear to be dispersed more widely within the suburbs than within the central cities. This pattern has been true of both black and white poverty families. In 1960, 59 percent of poor nonwhite families in the suburbs lived in concentrated poverty areas as defined by the Census Bureau, compared with 86 percent of the nonwhite poor in the cities.[27]

The black residents that newcomers to suburbia may find as their neighbors are somewhat more likely to consist of families with both husband and wife present than in the central cities. Estimates for 1969 indicate that 73 percent of Negro families in the suburbs have husband and wife in the home, compared with 67 percent in the central cities. In the suburbs, 24 percent of Negro families are headed by females, compared with 30 percent in the cities.[28] Although no comparative data are available on residential segregation by social class among blacks in suburbs and central cities, the limited housing choices open to blacks in the past have made it difficult for middle-class families to leave the central-city neighborhoods where low-income black families are concentrated. Thus, there is less class segregation among blacks in the cities than there is among whites.[29]

Racially integrated neighborhoods are difficult to find in both central cities and suburbs, and once again the differences are not great. If integrated nieghborhoods are defined simply on the basis of the racial mixture of the residents, many central-city areas that are in transition from white to black would be counted. In order to identify neighborhoods whose interracial character is likely to have some stability over time, one recent study uses as a criterion whether both whites and blacks are moving in. On the basis of a national survey, the authors estimate that, in 1967, 9.5 percent of black families in the central cities and 9.6 percent in the suburbs were living in "substantially integrated" neighborhoods—with more than 10 percent Negro fami-

[27]U.S. Bureau of the Census, *Poverty Areas in the 100 Largest Metropolitan Areas,* Report PC(S1)-54 (November 1967).

[28]U.S. Bureau of the Census, *Current Population Reports,* Series P-20, No. 200 (1970), p. 13.

[29]Karl F. Taeuber and Alma F. Taeuber, *Negroes in Cities* (Chicago: Aldine, 1965), p. 182.

lies and both Negroes and whites moving in.[30] A slightly higher proportion of suburban than central city blacks are estimated to be living in integrated neighborhoods where the black population is less than 10 percent; and these neighborhoods may well be the places that will offer good opportunities for racial integration in the future.

The suburbs have a clear advantage over the cities in their ability to offer living environments relatively free of crime. There is substantially less major crime of all kinds in the suburbs than in the central cities. The Department of Justice Uniform Crime Reports indicate an average of twice the rate of serious offenses per 100,000 population in the central cities as in the suburbs of the thirty-seven largest metropolitan areas. In the Northeast the difference is still greater, with central-city rates more than three times as high as those in the suburbs.[31]

In summary, then, the advantages of the suburbs in terms of housing and neighborhoods are less than they are usually presumed to be. Whether black families are likely to find housing of good quality in socially and racially mixed neighborhoods more easily in the suburbs than in the central cities is uncertain and depends upon where in the suburbs they go. They are, however, more likely to find single-family houses that they can buy in the suburbs, and they are more likely to be able to find a safe neighborhood in which to live.

What Government Can Do

The suburbs clearly could provide black people with improved access to a series of important resources: to better jobs, schools, housing, and neighborhoods than most blacks have in the central cities. Yet today's black residents of suburbia have not fared much better than those in the central cities, on the average, and the advantages they do have are limited and uneven. Any policy that hopes to make substantial reductions in racial inequality by encouraging more blacks to settle in the suburbs must come to grips with the fact that suburban settlement in the past has not contributed greatly to this goal.

Why the suburban potential has not been realized to a greater extent is puzzling. More information on who the blacks are in suburbia, where they came from, and what kinds of communities they now live in would help to explain what has been happening. Still more useful would be case studies of

---

[30] Seymour Sudman, Norman M. Bradburn, and Galen Gockel, "The Extent and Characteristics of Racially Integrated Housing in the United States," *Journal of Business of the University of Chicago*, 42 (January 1969): 50-92.

[31] U.S. Advisory Commission on Intergovernmental Relations, Bulletin 70-1, "Metropolitan Disparities," Table VII.

families over time, to see how they have made use of suburban institutions and opportunities and how these have affected their lives. Without such information, the most plausible explanation is that the suburbs have not been truly open to black newcomers—housing has been available only in certain limited locations, and suburban governments have not been very responsive to the needs of black residents. Many blacks who live outside the central cities are located in old, industrial communities in the North which have been engulfed by metropolitan expansion, and in rural shack towns in the South which are in the outer fringes of metropolitan counties. But even where these conditions do not hold, blacks living in suburbia have not necessarily been well integrated into its network of services and opportunities. A recent study of Nassau County, New York, found the poor residents of this generally wealthy suburb disadvantaged by virtue of their isolation in neighborhoods far from public services, by the lack of effective public transportation, and by inadequate information on how to negotiate the complex geographic and institutional environment.[32] This study recommends government action to develop low-income housing, improve public transportation, and provide day-care centers, training and job development, service centers in poverty areas, neighborhood health centers, and an information and referral system.

To espouse a policy of helping blacks leave the central cities without attending to where in suburbia they go or what happens to them after they arrive suggests a historic analogy. One might have argued fifty years ago that the key to advancing the economic and social status of Negroes was to help them leave the South. Jobs were expanding faster in the North and West, and school systems, public services, and housing conditions were all better than in the rural South. In the long run, migration out of the South has helped narrow the many gaps between blacks and whites, but the process has been slow, painful, and uneven, and fifty years later the gaps remain large even in the North and West. Migration from the South did not give the Negro wide choices of where to live in the North and West, nor did it deal with job discrimination, slum housing, and services and institutions unresponsive to the special needs of the migrants. In similar fashion, today's suburbs are by no means free of discrimination in hiring and promotion, nor do they offer wide choices of housing and neighborhoods to incoming blacks, nor have they advanced even as far as the hard-pressed central cities in developing services to match the needs of the black poor.

That a more rapid movement of black population to the suburbs is no panacea for the problems of racial inequality does not, of course, mean that it

---

[32] Organization for Social and Technological Innovation and Llewelyn-Davies, Banks, Forestier-Walker, and Bor, "Poverty in Spread City: A Study of Constraints on the Poor in Nassau County" (Cambridge and New York, 1969).

is unimportant or unworthy of government support. There are other grounds for supporting freer access to suburbia than the ones dealt with earlier in this discussion. Freedom of residential choice and equal access to housing markets would in themselves remove significant disadvantages under which blacks now operate. Blacks will not receive equal treatment in the housing market without government intervention to establish and protect this right. No other course of action is compatible with our national values, and I believe that no further justification is needed.

Public policy aimed at equalizing access to housing markets could appropriately concern itself with effective enforcement of civil rights laws, monitoring of real estate practices, and programs to alleviate some of the pains of racial transition in residential areas. But there is also a strong case for more positive government action in the suburbs, as part of a strategy for improving living conditions in the central-city areas where blacks are now concentrated. If the existing ghetto neighborhoods had to absorb much of the growth in black population, the programs already under way to equip them with new facilities and improved services would be virtually impossible to carry out. To provide space for new schools, recreation areas, health centers, and even a limited amount of new housing, these crowded neighborhoods will have to lose population. Unless there are positive programs to help some residents move elsewhere, the commitments that have already been made in model cities programs, urban renewal, and community action programs will lead to destructive conflicts over limited space, while already inadequate services become still more overburdened trying to serve larger populations. Most blacks who leave the crowded inner-city neighborhoods will probably continue moving to the outer areas of the central cities, but these neighborhoods themselves will be hard-pressed to maintain decent environmental conditions and services unless the suburbs also provide a share of the needed living space.

What forms of government action would most effectively provide better housing opportunities for black families in suburbia is not at all clear, and the answer varies from one community to another. The list of possible actions includes removing zoning and building code restrictions that raise the cost of new construction, developing or renovating subsidized housing for low- and moderate-income families, making housing subsidies available to private or nonprofit developers, supplementing the incomes of poverty families to increase their ability to pay for housing, and assuring an adequate flow of investment funds into home mortgages at reasonable interest rates.

Although the appropriate government measures are difficult to define, the governmental pattern most conducive to opening the suburbs is easier to identify. Recent experience reflects a governmental pattern that consistently obstructs the free movement of the inner-city poor and minorities to the suburbs. This is a pattern in which local suburban governments are given opportunities to decide on who shall be permitted to enter. Most current housing subsidy programs, as well as the assignment of responsibility for land development and housing regulation, offer the suburbs abundant opportunities to limit the amount of housing available to low-income families, either through acts of omission or commission. As long as the structure of government permits suburban communities to behave like country clubs screening new membership applications, public policy is not likely to make it easier for black people to move to the suburbs. Nor is this situation likely to change dramatically as the number of blacks already in suburbia increases. Recent reports in the *New York Times* indicate that middle-income blacks have been joining white families in attempts to keep subsidized low-income housing from being built in their neighborhoods in a series of communities across the country, including North Hempstead, Long Island; Flint, Michigan; Greensboro, North Carolina; Cleveland; and Philadelphia.[33] This is a pattern of opposition familiar from urban renewal controversies a decade ago, where racially integrated central city neighborhoods also fought against low-income housing, giving rise to the slogan, "Black and white, shoulder-to-shoulder, against the working class."

Within the present pattern of suburban government, the best prospects for blacks to move to and within suburbia will continue to be those created by the private housing market. When existing houses are available at moderate cost and black families know of their availability, racial change is not likely to become a political issue and, in any event, there is little that local government can do to prevent blacks from buying homes they can afford. However, suburban communities may attempt to remove cheap housing through urban renewal programs or through strategic planning of new highway routes. Housing subsidy programs will have the best prospects of success in the suburbs if, as is the case with the new federal programs for homeownership and rental housing, they operate through private developers (including nonprofit sponsors) rather than through agencies of local government. Housing allowances, which would make subsidies available to families directly and permit them to buy or rent existing housing, would be even more effective. Alternately, subsidies administered through higher levels of government, such as state housing agencies, may also be able to operate effectively.

If the provision of services designed for low-income residents is also left in the hands of suburban localities, the supportive role of government will probably continue to be very limited in the suburbs. As black populations increase, especially as the number of poor families grows, suburban communities will suddenly discover fresh problems in their midst. They are likely to

---

[33]"Blacks on L. I. Kill Plan to House Poor Blacks," *New York Times*, July 24, 1970, p. 27.

take some action in response to political pressures or simply to newly recognized needs. But middle-class suburbanites until now have not shown themselves eager to pay higher local taxes to support new services or to remake their local institutions so that they are more in tune with the needs of the poor.

A new suburban pattern may well emerge in which large numbers of poor people will live in locations scattered across a wide range of communities, constituting only a small minority in each place. With little political or economic power available to them, the suburban poor may very well be unable to command an equitable share of the conventional public services, let alone persuade their local governments to provide special services for which middle-income families have little need. This prospect argues strongly for state or metropolitan attention to the distribution of public services and particularly to the provision of poverty-linked services. Local governments are likely to provide very few of the latter, partly out of fear that to do so would attract more poor people into the community; whereas higher levels of government could act without undue concern about migration from one locality to another. Also, the dispersal of poor people throughout suburbia could very well leave many communities with too small a number of eligible families to provide an efficient scale of service.

The pattern of government that would be most helpful to the black poor would be one in which higher levels of government either provide, or finance, poverty-linked services. When the states administer welfare and health services on a uniform basis, low-income families are least likely to be disadvantaged by moving to the suburbs. As the suburbs become increasingly aware of the problems of poverty within their own territory, the prospects for building political coalitions with the inner cities to press for more state and federal aid should improve. The *New York Times* recently interviewed political leaders in the suburban counties around New York on their willingness to help provide jobs and housing for the inner-city poor. They responded uniformly by pointing to poverty problems within their own counties, and by calling for more federal aid for low-income housing, public transportation, welfare, and medical care.[34]

The suburban lag in providing poverty-related services, together with the fact that the white and black poor are already on the increase in suburbia, argue for giving early attention to improving these services. Once the suburban setting itself becomes better equipped to help the poor advance, more rapid movement of blacks into the suburbs may contribute more noticeably toward ending racial disparities.

---

[34] "Suburbia Cool to Agnew Advice," *New York Times*, March 9, 1970, p. 22.

3 Litigation and the Dynamics of Unanticipated Change

DANIEL WM. FESSLER*

If the preceding papers with their focus upon the chronic physical infirmities and spiritual disaffection of contemporary "urbanitis" have left the reader with the impression that a typical metropolitan governance scheme is an accident awaiting a location, it remains for this monograph to chart some dangerous intersections. At this late hour in the governance crisis, that combination of determined and passive force which sustains the status quo is threatened with assault by an ancient, but heretofore little-used weapon—litigation in the hands of politically and economically disadvantaged. As an alternative to the ad hoc political alliances necessary to the formation of even a temporarily effective "crisis constituency," this appeal to a nonmajoritarian tribunal offers both speed and relative precision to groups which, to this hour, have been either noncombatants or inneffective forces. The following pages will explore both the origin and dynamics of this strategy, and offer suggestions regarding its potential impact.

The concept of taking "city hall" to the county courthouse has never gained broad popular usage in American legal practice. A diversity of factors accounts for this sparse record. Until fairly recently, the institutions of local government were insulated from judicial scrutiny by a vestige of the royal prerogative—sovereign immunity. Equally disabling was a rather stringent judicial attitude toward the qualification of an individual citizen to call into question the exercise or processes of executive or legislative decision making. The lawyer's term for this jurisdictional difficulty is "standing." Both "sovereign immunity" and "standing" are doctrinal manifestations of a fundamental

*Acting Professor of Law, The Law School, University of California at Davis.

attitudinal limitation on the power of the judiciary as a coordinate branch of government.[1] Yet perhaps more decisive than any popular attitude in curtailing the pressure of the judicial process upon both the personnel and institutions of local government was the absence of an interested segment of the Bar. The very individuals or groups who found themselves without economic or political power were the least likely to obtain the "for hire" services of the private Bar. The magnitude of the potential defendant, when contrasted with that of the citizen-plaintiff, combined with the lack of favorable procedural or substantive precedent to dissuade all but the most unusual case from ever seeking a day in court.

Today, with what may strike some as alarming speed, the theoretical barriers to judicial redress seem to have slipped, if not fallen, and the ranks of advocates have been populated by an aggressive and ambitious group which evidences little inhibition about invoking the processes of either state or federal courts.

Years from now, when the developments of this movement are the object of disinterested history, it may be concluded that the decisive clash between an activist judiciary and the institutions and personnel of local government began with the civil rights cases in the early 1950's. These first public accommodations and school desegregation cases set a pattern which, for better or for worse, has structured the litigation campaign. With few exceptions, the law reform initiative has been declined by state courts. Thus pressure has concentrated upon the federal judiciary, and for the purpose of securing federal jurisdiction the controversy has almost invariably been elevated to the plateau of federal constitutional questions. Starting with *Brown* v. *Board of Education*,[2] the United States Supreme Court has impressed the lower federal judiciary in a most broad-ranging assault upon the segregated facilities and practices of a thousand units of local government.

In its quest to reduce race to a "Constitutional irrelevancy," the Court has swept away doubts as to the amenability to suit of either the personnel or institutions of state and local government. All were "persons" within the broad jurisdiction of the nineteenth century Civil Rights Act. More importantly, the Court extended the "thou shalt" and "thou shall not" remedy of injunctive relief to specifically command these constitutionally subordinate entities of government at the instance of citizen suitors. As most recent experience has demonstrated, these incidents of judicial activism carry the momentum of logical expansion. The rising level of individual expectation has created the demand that logic be pursued.

---

[1] Many judges have, themselves, resisted invitations to enter the "political thicket." This judicial reticence is summed under the abstinence doctrine termed "justiciability."

[2] *Brown* v. *Board of Education of Toledo*, 347 U.S. 483 (1954).

Yet the reversal of restrictive precedent, even the slow conversion of judicial attitude, would—in the absence of advocates—be ineffective to secure a massive intervention by the passive forces of the "third branch." The critical factor which may render unique this era in the relationship between the individual and the social collectivity is the "Law Reform Movement." And because an increasing tendency of this potentially irresistible force is to test the immovability of the object of local government, the origin of the "movement" is of more than passing interest to those who would seek to understand and to harness the forces so recently released.

Benefit of Counsel for the Poor

A consideration of the question of legal services to the disadvantaged, and the anticipated future conflicts with the harm-dealing or tolerating institutions that dot the governance landscape, cannot be developed in the absence of some basic data concerning the qualitative and quantitative problems faced by the economically and politically powerless citizens who are forced to suffer without benefit of counsel or representation a disproportionately oppressive number of "legal entanglements." In his presidential message to the organized Bar, published in the July 1970 issue of the American Bar Association *Journal*, Bernard Segal noted that there are in America today more than 20 million citizens who will endure legal problems without professional counsel or representation in the absence of some form of legal aid. While the message is marked with high praise for the efforts of both the organized Bar and concerned governmental officials, there can be no denying the fact that only a small portion of those citizens falling within the defined need for legal services are receiving this vital help. In its most immediate sense, the "problem" is that the poor need legal representation. To use President Segal's words, the task is to remedy the plight of those "millions of people [who] are without legal representation, when they suffer the sanction of the law but have no access to its remedies." At first glance it might be assumed that President Segal's thoughts were limited to the provision of representation in "private," e.g., nongovernmental, disputes. That potentially comforting thought was quickly dispelled as Mr. Segal went on to charge the Bar with

> . . . failure to become involved in the difficult and volatile problems which so gravely threaten our society: problems of the ghetto and the slum, of poverty and discrimination in our inner cities, of crime which threatens our safety and terrifies our people, of an environment which poisons our air and water and threatens the quality of life in America.

Any movement commissioned to grapple with these problems will, of necessity, confront the existing institutions of metropolitan governance. Yet reform within the ranks of the organized Bar when, and if, it materializes will find the aggressive forces of the legal service movement already positioned in the field. It is this movement that increasingly disturbs the tranquility of the status quo. It is quite new.

The records suggest that the first legal aid society in America was founded in 1876 to provide legal assistance to immigrants. From its inception, the movement was an urban phenomenon. Before the end of that century, legal aid societies were initiated in a number of large cities. Although a number were devoted to the needy of a particular nationality, there began to emerge a concept of a community-serving organization, chartered under auspices comparable to those of a modern United Fund. Reflecting the then prevalent attitude toward public assistance, the community legal aid societies rendered legal advice and representation to the urban poor as a charitable service. By the 1920's, the legal aid movement had attracted considerable support from within the ranks of the organized Bar. In 1920, Charles Evans Hughes delivered a landmark address to the American Bar Association in which he bore down on the obligation of the Bar to sponsor and to support legal aid to those who were unable because of material circumstances to compete in the marketplace for traditional legal representation. With Hughes as its initial chairman, a standing Special Committee on Legal Aid was created within the institutional framework of the American Bar Association.

What was the record of the next 45 years, a period characterized by local initiative and support? By 1965, there were extant 252 legal aid offices, providing legal services in civil matters, and 136 public defender offices, dealing with criminal matters. Overwhelmingly, these local efforts were in larger metropolitan settings. Some legal aid offices were manned by salaried staff, but the typical operation functioned with part-time volunteers who attempted to steer prospective clients in the direction of those members of the local bar who shared Mr. Chief Justice Hughes's concept of a lawyer's professional responsibility. The most credible figures record that legal aid societies provided representation for approximately 426,000 civil matters during 1965 at a cost of $5.4 million. The typical society was miserably underfinanced, invariably buried under a massive caseload, and the source of grossly noncompetitive salaries. Such an operation was incapable of a genuine law reform effort.

OEO's Legal Services Program

A dramatic shift in the heretofore sporadic governmental involvement was achieved with the advent of the Office of Economic Opportunity (OEO). Recognizing both the pressing need for legal services, and the fact that nearly all metropolitan governments had eschewed any opportunity to assist their local societies, the Legal Services Program was initiated as one of OEO's major Community Action Programs. This shift from local initiative to a federally supported and coordinated program may ultimately prove to be one of the central forces in bringing down the nonviable institutions of fragmented local government.

This massive infusion of federal support was designed to maximize the more successful models of legal aid operations. In those cities where the legal aid program was comparatively vigorous and successful, the program leaders were natural candidates for an OEO grant. In the majority of cases, grant applications were accepted from members of the private bar who were obliged to form nonprofit corporations that conformed to clearly defined governmental guidelines both for the structuring of the grantee organization and for the legal services it was to offer. A uniform financing scheme was adopted under which 80 percent of a local project's budget was to be assumed by OEO, and the remaining 20 percent share to be raised locally in the form of cash or donated services, facilities, and equipment. For reasons that reflected both political concern and the existing legal aid/public defender division, OEO Legal Service Projects were funded for counsel and representation in *civil matters only.* Yet having observed this fundamental division, the client services were defined so as to cover the full range of noncriminal matters, including administrative practice and basic law reform litigation. The hallmarks of the OEO Legal Services concept are characterized in the project requirements which have structured the current model:[3]

1. Projects were required to decentralize their operations and to open easily accessible neighborhood law offices in areas of client residential concentration. [Thus the concept of neighborhood legal services.]

2. Projects were given the mandate to engage in a broad program aimed at reforming both civil laws and administrative practices that adversely affected the poor.

[3]Although they reflect adherence to these criteria, the size and number of grantee projects seem to vary widely. Pre-OEO experience with legal aid and population distribution patterns appear to have been primary influences in determining the number of grantees within a given territorial or political entity. In some large cities (e. g., Denver) a metrowide legal aid society has become the grantee project with a multiplicity of neighborhood offices. In others (e. g., Boston) the grantees are fractionated along the lines of historical divisions of a problematically joined metropolitan community. In California, individual grantee projects service the metropolitan centers, whereas the rural areas and smaller communities are serviced by a vigorous statewide grantee, California Rural Legal Assistance Foundation (CRLA). Montana seems to be unique in that both the metropolitan and rural areas are served by branches of a single statewide grantee (Montana Legal Services Association). With the exception of several rural-oriented grantees and a small number of special efforts directed toward population concentrations of American Indians, the scope of the OEO Legal Services effort is essentially directed to metropolitan populations.

3. Insofar as it was possible, the traditional model of the lawyer-client relationship was to be retained, and to this end individual clients were to receive high quality legal representation in every case.

4. The grantee project was charged with the development of a law reform effort that would contribute to the economic development of the community served by the provision of legal skill to local enterprises.

5. As developed by the project, the total services program was to include efforts at preventative legal education.

6. In order to preserve the integrity of the neighborhood community, the OEO-funded legal services project could provide advice and representation to organized groups in the community served.

The model thus developed—a community-based, community-promoting services project—is of particular interest given the Committee for Economic Development's recommendations for a two-tiered scheme of metropolitan governance.[4]

There is an additional dimension of the current OEO Legal Services Program which is worthy of brief mention. Recognizing that in their disproportionately heavy incidence, if not in their basic nature, the legal problems of the disadvantaged are unique, OEO has sought to buttress the quality and quantity of the practice conducted by the various legal service projects with "back-up centers," charged with the development and projection of both substantive and procedural innovations which the projects can advance for local legislative or judicial adoption. As of this writing, OEO-sponsored back-up centers specialize in health law (Los Angeles); consumer law (Boston); housing and economic development (Berkeley); welfare law and policy (New York); juvenile law (St. Louis); and education (Boston). In each instance, these research and policy formulation centers have their headquarters at universities with the incidental benefit that the curricula of the immediate (and ultimately remote) schools have been enriched by the accumulated expertise in an aspect of "poverty law," and designed to sensitize students to their obligations to, and the problems of, the poor. In addition to these specialized centers, a delivery and coordination system by which reform proposals and shared experience can be disseminated to diverse attorneys in the field has been instituted at Northwestern University in the presence of the National

[4]As used in this context, it should be obvious that "community" would bear only a coincidental identity with a metrowide population. The emphasis is on the subgroup—the neighborhood—and the conception of interest there developed.

Entitled "Reshaping Government in Metropolitan Areas," the statement by the Research and Policy Committee of the Committee for Economic Development was released in February 1970. Whether the concentration of litigation weapons in the hands of the diverse immediate levels of organization is at war with the concept of shared powers remains to be seen.

Clearinghouse. This combination of neighborhood legal services projects and substantive law centers affiliated with urban universities has given reality to a "poverty law movement."

As previously explained, the pre-OEO legal aid scheme, as of 1965, was meeting less than 2 percent of the legal needs of the disadvantaged. What more has been accomplished to date? The most recent figures clearly show that substantial improvement has been achieved in the less-than-7-year history of OEO Legal Services. As of this writing, OEO funding supports, in whole or in part, more than 850 legal services offices in 49 states, the District of Columbia, Puerto Rico, and the Virgin Islands. These offices are staffed by nearly 2,300 full-time attorneys. In contrast to the estimated expenditure of $5.4 million by the total of all legal aid operations in 1965, the budget for OEO Legal Services in fiscal 1969 was $58 million. This increased level of appropriations was substantially preserved in fiscal 1970.

Notwithstanding this period of rapid expansion, the record of accomplishment by both the organized Bar and government at all levels remains decidedly unimpressive when measured against the extent of unmet need. In marking the fiftieth anniversary of the founding of the Special Committee of Legal Aid Work, ABA President Segal cautioned that before congratulations are due on the progress of the past 5 years, we must recall that *less than one-fifth of the defined need for legal services by the nearly 20 million citizens deemed currently eligible is being met by all efforts, both private and governmental.*

Furthermore, we cannot be certain of the continued existence of such a legal services movement. Early developments within the Nixon Administration cast substantial doubt as to whether the Office of Economic Opportunity model, as inherited from the previous administration, would retain its viability. The actual record of attempts to materially weaken the legal services program should give rise to cautious optimism. The Green and Murphy Amendments, which would have shifted the ultimate control over programs from federal to state officials, were defeated with the active opposition of the organized Bar. An Administration-sanctioned "decentralization plan" was withdrawn by the then Director of the Office of Economic Opportunity in the face of mounting opposition from the legal community. In short, the independence and professional integrity of the legal services movement seem to command the active attention of the Bar which, when fused with vocal support from the increasingly militant ranks of the client group, produces a powerful lobby.

Aside from the strength of this political coalition, two other factors deserve brief mention. In a collateral development that may engender short-run stability to the status quo, the President's State of the Union address in January, 1971, called for the abolition of a number of current cabinet-level departments in a fundamental restructuring of the executive branch. The fate of such a sweeping proposal is uncertain, and yet there can be little doubt but that the political struggle will be protracted. This condition could be of material benefit to the continued status of OEO Legal Services. For several years the Office of Economic Opportunity has felt threatened by shifts of its operative programs into older, cabinet-level departments. There have been vocal suggestions that Legal Services—now that it has advanced beyond the pilot program stage—should be shifted to either the Justice Department or the Department of Health, Education, and Welfare (HEW). With the future structure of these organizations now in question, the transfer or restructuring of the Office of Economic Opportunity may be viewed as a very subordinated project.

The current activities of HEW, and to a lesser degree those of the Departments of Housing and Urban Development (HUD) and Transportation (DOT), indicate support for the OEO Legal Services model. As of this writing (1971), HEW funds five research and demonstration grant projects, which have adopted the full-time independent poverty lawyer concept but have attempted to extend these services to the rural poor. Two of the HEW grant programs have experimented with the "judicare" concept whereby legal services for the poor are purchased directly from members of the private Bar. In August 1970, the Administrator of the Social and Rehabilitation Services Administration commissioned a study of the HEW future in legal services predicated upon the experience derived from the research and demonstration projects. The guideline furnished to that study group expressly predicated a future HEW role upon the continued existence of the OEO Legal Services Program, and envisioned a supporting, or resource-contributing, function for HEW. In the last two years HUD has funded at least two neighborhood legal services offices in Model Cities Target Areas. Again, support for the OEO program was given by the use of the OEO grantee as the recipient of these expanded federal monies. The Department of Transportation is also said to be giving active study to the use of neighborhood legal services programs which would specialize in the problems created by the relocation of transportation arteries.

If one were to attempt to discern a trend from these diverse factors, it would be that there appears to be a high level of satisfaction with the basic legal services model—a full-time poverty group advocate functioning at the neighborhood level without reliance upon local funding and enjoying the independence from the sanction of local political retaliation which is ensured by federal fiscal support and the watchful protection of a nationally organized Bar. For those concerned with the evolving processes and personnel forming the metropolitan governance structure the lesson is clear: *The presence of an advocate for the politically powerless seems permanent, and the occasion for resort to the machinery of judicial decision making is presumably in the future of nearly every governance crisis.*

The Experience with Litigation

The discussion that follows will seek to make several observations concerning the role of litigation as a stress-dealing and relieving factor in a governance crisis. First, litigation, or the threat of litigation, provides an immediate weapon with which the politically and economically disadvantaged may seek to influence the institutions and officials of local government. Second, the dynamics of the litigation process almost ensure that recourse will be had to external value systems and criteria in the process of deciding the issues of equality and minimum entitlement. Thus, not only is the potential of litigation that of a "disruptive" influence or pressure—when viewed from the vantage point of the proponents of a status quo or the "power structure"— but it is inherently at war with "local" solution, as that term is used to express the common notion that a settlement should reflect the alignment and realignment of pressures and interests at the purely local level. Even a "local" judge must decide a legal point with reference to the external criteria reflected in prior case law, and of course his decision is subject to appellate review by a tribunal of "out of towners." These observations suggest that litigation under either federal (Constitutional) or local (common law) theories offers an alternative way to bring to a crisis stage the demand for some relief on behalf of those who hold, as a matter of birthright or repetitive circumstance, the "short end of the stick" on the "wrong side of the tracks." To this previously neglected client group, the presence of neighborhood legal services attorneys offers potent weapons in the speed with which litigation may be used to focus and center a crisis and the ability of the judiciary to override "local" arrangements by placing an existing patchwork system under the strain of a decree of citizen entitlement, which can only be satisfied by a reordering of the governance structure.

Already, these lessons have been reduced to experience. To date, the skirmishes with legal services advocates have received their greatest attention in the fields of public assistance and education. Through a series of suits, which culminated in an assault upon the New York City welfare system before the United States Supreme Court, the procedural norms for the administration of welfare programs were removed from the hands of local decision makers and recast in the mold of a Constitutionally based articulation of procedural Due Process, binding upon every level of government. Although less effective in attacks upon the substantive question of "what" may be Constitutionally claimed as a matter of entitlement from state and local government, the success of the more or less coordinated attacks upon state and local public assistance practices has been a major force in nationalizing what was regarded two generations ago as a matter of local charity.

In the field of education, the early preoccupation with a broad-ranging implementation of the desegregation decrees has given way to the assertion of claims more immediately identified with neighborhood communities. Thus, current litigation efforts, in addition to seeking "community control" over the neighborhood schools, attempt to secure special instructional programs geared to ethnic or racial backgrounds. The emergence of subgroup power is most dramatically evidenced in suits such as the one being prosecuted by the Chinatown Office of San Francisco Neighborhood Legal Services. In *Lau* v. *Nichols* the United States District Court for the Northern District of California is confronted by a class action brought by Chinese-speaking students who allege that defendants' (San Francisco Board of Education members) failure to provide special instruction in English violates rights guaranteed under the Fifth, Tenth, and Fourteenth Amendments to the U.S. Constitution, under designated sections of the California Constitution and State Education Code, and under the 1964 Civil Rights Act.[5]

Litigation in the education and public assistance cases has been the source of recurrent clashes between neighborhood advocates and the institutions of metropolitan governance, but it is in the areas of housing, transportation, land redevelopment, and municipal services that the disputes have become increasingly numerous and sharply focused.

That the disadvantaged bear the most oppressive burden of blighted housing, inadequate transportation, and inferior governmental services has been documented in several of the related monographs in this study. That these conditions are increasingly the target of remedial litigation efforts should thus come as no surprise. The potential role of proffered federal assistance to the faltering cities forms a primary target in this poverty-law strategy. Thus, within the last year the San Francisco Workable Program was rejected by the Department of Housing and Urban Development at the instance of the "Citizens' Emergency Task Force for a Workable Housing Policy of the City of San Francisco." Members of the "West End Community Conference" have brought suit against the St. Louis Land Clearance for Redevelopment Authority seeking the same objective.[6] In these representative cases the disadvantaged were able, with relative dispatch, to secure judicial assessment of *their own grievances* in a forum that did not require that their position be diluted

---

[5] *Lau* v. *Nichols*, No. C70–627LHB (D. Cal., March 23, 1970).

[6] *St. Louis West End Community Conference* v. *St. Louis Land Clearance for Redevelopment Authority*, No. 70C71(3) (E.D. Mo., filed Feb. 13, 1970). Highway relocation plans in Austin, Texas, were the object of a suit filed by neighborhood legal services attorneys in *Concerned Citizens for the Preservation of Clarksville* v. *Volpe* (U.S.D.Ct., W.D. Tax. 1970).

by participation in a "crisis constituency." Perhaps the term "crisis constituency" requires clarification. We begin with the thesis that within a metropolitan population numerous subgroups may be identified. Some are ethnic; others organize around common religious, social, or economic ties. Each of these groups is the source of articulated disaffection with "city hall." To one, the problem is a lack of control over the neighborhood school. To another, the key issue is a perceived inadequacy of police protection; to a third, the problem lies in a perception of an oppressive incidence of threatening police patrols. If recourse is had to the electoral process, it is obvious that no one of these groups will command a sufficient following to replace major incumbent officeholders. To accomplish this goal a majoritarian constituency may be gathered and held in concert for the duration of an election campaign. It is in the dynamics of mixing and matching subgroup goals that the "crisis constituency" is born. United only in a desire to "throw the bastards out," the crisis constituency is a temporary gloss, fusing elements that are intrinsically at odds. True discussion of the issues is necessarily blurred, for if accurately portrayed, the issues would destroy the ad hoc alliance! If successful in the short-term goal, the constituency immediately dissipates and a process of incremental disaffection begins as the new officeholders experience the painful reality that they cannot serve a host of discordant masters.

More often, the crisis constituency is defeated or repelled. This, too, can be explained. Because it is united by only the thinnest veneer of unison, the ranks of the constituency are subject to depletion as the incumbent officials move deftly to satisfy a sufficient quantum of subgroup demands to neutralize that particular neighborhood, synagogue, or ladies auxiliary. This accomplished, the remaining forces can be handily defeated on election Tuesday. In such a process the economic and social inconsequentials are inevitably among the ranks of the unsatisfied. Unable to bargain, they have no formulated "price" to demand as an inducement to abandon the crisis constituency. Though nonedifying in the description, this is rather clearly the "American way." Upon further reflection, this homely process emerges as a functioning variety of democratic rule.

Standing in sharp contrast are the dynamics of judicially wrought reform. The long-range impact of this ability to "go it alone" should prove of great significance in the future of metrowide planning. It is not too early to suggest that one of the dangers involved in this type of litigation is the fact that when a particular subgroup attacks a general governmental plan or policy, the judicial review is made without the participation of constituencies whose interests were "traded off" so that central planning might be accomplished. Whether the danger that the court will miss the "big picture" should outweigh the desirability of its ability to relieve against the rigors of a particularly sharp exaction is an ultimate policy question. A balancing test whereby the specific harm complained of is weighted against the general good sought to be accomplished (discounted by the availability of less onerous alternatives) is emerging as the test for judicial review.

Urban land-use questions frequently involve the most naked instances of struggle between subgroups within the metropolis. Since urban land is a good of obvious scarcity, land-use decisions are among the most vital ones committed to "local government." Ligitation in this area is conducted on two fronts. On the first, it seeks to enforce existing ordinances or regulations for the benefit of legal service clients, while on the second, litigation seeks to accomplish heretofore unarticulated reform.

Not infrequently, the "legal" interests of the ghetto resident have been met by the passage of local ordinances which, once the particular crisis constituency has dissipated, are the subject of distressingly lax enforcement. When a reading of the local ordinance books reveals such a condition, the task of the legal services lawyer has been to demonstrate that for his clients the "law is not dead, but merely sleeping." An example of such direct action may be found in the recent suit to compel the New York City Department of Housing to exercise its statutory powers to require that landlords adhere to the City Building Code.[7] Federally assisted housing programs are generally attended by similar benign provisions that are either unappreciated or ignored by local officials. In *Cooper* v. *Cook County Housing Authority*, the plaintiffs, black tenants in public housing projects, launched a direct attack upon the local political "truce" with their suit in a United States District Court which sought to enforce local compliance with Housing and Urban Development Departmental guidelines for project site selection which require that members of minority groups be "given an opportunity to locate outside of areas of concentration of their own minority group."[8]

[7]*Moultrie* v. *Mayor of City of New York*, 163 N.Y.L.J. No. 70 (N.Y.S.Ct. 1970).

[8]*Cooper* v. *Cook County Housing Authority*, No. 70C595 (N.D. Ill. 1970). The truce between the federally supported Chicago Housing Authority and Mayor Daley's political machine regarding the selection of sites for public housing projects had been dealt a sound blow one year earlier. In *Gautreaux* v. *Chicago Housing Authority*, Civ. No. 66C1459 (N.D. Ill. July 1, 1969), a United States District Judge at the instance of impoverished black suitors had prescribed in detail the steps that must be taken "to prohibit the future use and to remedy past effects of the defendant Chicago Housing Authority's unconstitutional site selection and tenant assignment procedures . . . . " Under the "political arrangement" which had heretofore obtained in Chicago, the CHA had agreed to reject a proposed site because the alderman in whose jurisdiction the project was to be located reported that members of his constituency were opposed to the racial and economic integration the project might bring. Under this practice the Housing Authority had placed nearly all of the housing projects in black neighborhoods thus reenforcing the pattern of racial segregation and incidentally working a de facto exclusion of nearly 188,000 white families who were eligible for public housing but who elected to forego the opportunity rather than move into all-Negro projects in all-Negro neighborhoods.

Yet as disquieting as these attempts to put substantive sacrifice behind what had heretofore been empty gestures, litigation seeking to enforce local ordinances or administrative guidelines pales when contrasted to suits grounded in an assertion of Constitutional rights. Again, the housing and land-use areas are replete with examples of this type of litigation. In *Cole v. Housing Authority of Newport*, the Rhode Island Legal Services Corporation was successful in obtaining a federal decree rendering null and void a locally imposed two-year residency requirement for admission to public housing in the City of Newport.[9] The United States District Court rested its order upon a determination that such a local residency requirement violated the Fourteenth Amendment's Equal Protection Clause.

The Precedent Value of Poverty Law

When a court rests its judgment upon an interpretation of the federal Constitution, that judgment becomes a rule of decision which rises to strike down similar programs or policies wherever they might be practiced! This dynamic of the litigation process requires some explanation. If the rule of decision involves an interpretation of rights arising out of a local ordinance, such a judgment is incapable of application (save by analogy) in situations where that particular ordinance is not present. Here is an example of a truly "local ruling." If one advances beyond that case to postulate a suit in which a state statute was relied upon as the ground of decision, an application of that statute to a factual situation within that state creates a judicial precedent which, if affirmed by the Supreme Court of that state, has made "law," not only for that locality but for all other localities within that jurisdiction! But there is more. Where the ground for decision is an application of a federal statute, the precedent value (meaning a rule that will oblige inferior courts in the decision of future cases involving the same statute) is of national significance and obligation.

Yet in all of the precedent examples to this point the propounded judicial relief is still subject to prospective reversal at the hands of legislative enactments cast at that level of government of coordinate dignity with the pronouncing court; for example, a city council may alter or repeal an ordinance; the state legislature may nullify the provision of the state statutes relied upon as a ground of decision; and, in the instance of a federal statute, the Congress retains a plenary power to amend the rights so fixed and declared.[10] But where the ground relied upon by a court is a provision of the

---

[9]*Cole* v. *Housing Authority of Newport*, Civ. No. 4265 (D. R.I. Jan. 27, 1970).

[10]Where the ground of judicial decision is an interpretation or application of an administrative norm, such as a requirement of the Department of Housing and Urban Development enunciated by the Secretary pursuant to rule-making powers delegated by

federal Constitution, the rights and obligations so declared are fixed—for practical purposes—beyond the power of alteration by any institution save for that court or an appellate judicial tribunal. When the United States Supreme Court affirms such a decree, as in the decision that the regulations of the City of New York's public assistance program were Constitutionally deficient in their failure to grant pre-termination or suspension of evidentiary hearings, that rule of decision automatically declares the point not only for the City of New York, but for the Department of Health, Education, and Welfare, and every public assistance program in the United States! Only the volume of publicity that followed *Goldberg* v. *Kelly* prevented this mandate from being the source of the essence of unanticipated change![11]

Indeed, the "federalization" of the "welfare problem" may be accurately attributed to the impact of three key suits which sought to establish the Constitutional invalidity of specific facets of designated state or local programs, which were, in turn, instances of procedures or policies common to nearly all state and local administrations. The cumulative impact of these· decisions has been to escalate the public assistance recipient's "level of entitlement" to the point that there is an emerging political consensus that Constitutionally acceptable programs have been "priced" beyond the ability of state and local government.[12]

Congress, it is generally accurate to say that the pronouncing agency retains the power to alter or amend the regulation.

[11] The foregoing analysis suffers from a technical impairment in its suggestion that as a result of the Supreme Court's judgment in the New York City welfare regulations case, the administrative regulations of any public assistance program which failed to provide for a prior hearing were automatically invalidated. They were to the extent that in any judicial attack brought at the instance of local suitors any court would be obliged to declare them fatally deficient. Yet the mandate would require for its specific implementation the formality of a local suit. Most jurisdictions, realizing the futility of resisting such a decree will so alter or repeal the offending local practice or regulation as to make the local litigation unnecessary.

The essential point is thus quite accurate: in the course of passing upon what would appear to be an individual case, the United States Supreme Court's interpretation of a Constitutional requirement settles the question beyond the capacity of any institution of government to change it, save for a Constitutional Amendment or a subsequent decision of that Court.

[12] In chronological order these decisions of the United States Supreme Court are: *Smith* v. *King*, 392 U.S. 309 (1968) (Alabama's "Man in the House" regulation contravenes the Social Security Act); *Shapiro* v. *Thompson*, 394 U.S. 618 (1969) (Pennsylvania residence requirement for receipt of public assistance violates Equal Protection guarantees of the Fourteenth Amendment); and *Goldberg* v. *Kelly*, 397 U.S. 254 (1970) (New York City's failure to accord welfare recipients an evidentiary hearing before suspension or termination of an established grant violates the Equal Protection Clause).

Not all such cases were successful. In *Dandridge* v. *Williams*, 397 U.S. 471 (1970), the Court refused to invalidate a provision of the Maryland Assistance Program that fixed a

As a parenthetical observation, it is worth mentioning that the history of public assistance reform reveals a striking instance of litigation in the vanguard of the movement, with large-scale political organization arising on the strength of expectations created in its wake. Thus the National Welfare Rights Organization, which is currently the sponsor of additional reform litigation, followed OEO Legal Services. One of the dynamics of the movement is that noted success in one jurisdiction seems to produce immediate demand for a "local replay." The precedent value of the initial decision receives part of its practical implementation from popular pressures.

The critical lesson bearing repeated emphasis is that, for the institutions of local government, changes in legal restraints and enforceable obligations can arise in the context of litigation to which the particular institution is a total stranger. Indeed, at the local level, there may be complete ignorance of the fact that change is contemplated until that change has already been legally consummated beyond the ability of any combination of local forces or immediate majorities to amend or recall!

The thoughtful reader will perceive a major limitation upon the precedent pattern just described. In order for the "foreign" precedent to apply, the local situation must feature substantially the same rule or condition as was struck down in the initial suit. Thus, by precedent extension, the *Kelly* case invalidated the administrative hearing provisions of *all* public assistance programs to the extent that they failed to accord pretermination review. Beyond this major change, however, the precedent had no direct application to substantially analogous problems in related factual contexts. After *Kelly*, could the administrator of a public housing project evict a tenant family without a prior hearing? In seeking the answer to this very practical question, we encounter what is perhaps the least understood and most expansive dynamic of the litigation strategy—the power of legal *analogy*.

To understand the lawyer's fascination with the use of analogy is to grasp a significant insight into the function of the law. The late Mr. Justice Holmes described this phenomenon in the context of another problem in another era; yet his analysis is unmatched both as a precise expression and as an understanding exposition of this process.

... The training of lawyers is a training in logic. The processes of analogy, discrimination, and deduction are those in which they are most at home. The language of judicial decision is mainly the language of logic. And the logical method and form flatter that longing for certainty and for repose which is in every human mind. But certainty generally is an illusion, and repose is not the destiny of man. Behind the logical form lies a

---

ceiling on the maximum welfare grant, notwithstanding the family need computed by the number of dependent children in the recipient's home.

judgment as to the relative worth and importance of competing legislative grounds, often an inarticulate and unconscious judgment, it is true, and yet the very root and nerve of the whole proceeding. You can give any conclusion a logical form. You can always imply a condition in a contract. But why do you imply it? It is because of some belief as to the practice of the community or of a class, or because of some opinion as to policy, or, in short, because of some attitude of yours upon a matter not capable of exact quantitative measurement, and therefore not capable of founding exact logical conclusions. . . .[13]

Given Holmes' perspective, it can be understood why significant progress in the vindication of the legal rights of the disadvantaged had to await the arrival of "poverty lawyers." Until the "poverty clients' " conception of the "relative worth and importance of competing legislative grounds" could find its reflection in a lawyer's mind, judicially enunciated "law" was substantially beyond their reach. The poverty-law, or legal services, movement is a revolution precisely because it significantly expands the "belief[s] as to the practice of the community or of a class" and the "opinion[s] as to policy" which now must be weighed and accounted in the evolution of those compromises or trade-offs that are the object and exercise of governance. At the level of local government, this means that decision makers will be forced to respond to the sometimes shrill and unreasonably self-centered voices of a constituency which has heretofore only whispered, or suffered to have its needs and desires made the object of afterthought and more-or-less benign assumption.

Poverty Law and Public Services

Again, the significance of this potential is perhaps best illustrated by contemporary example. Even the casual observer of metropolitan life in this most prosperous nation is struck by the vast disparities in both the quantity and quality of municipal services. Indeed, so pronounced are these inequalities in the levels and kinds of services received from local government that the cliche "the wrong side of the tracks" has become, through three generations of usage, a designation understood by all. To attack this condition through litigation—to equalize the "immediate environment"—is perhaps the ultimate clash between the neighborhood legal services advocate and the institutions of metropolitan governance. The initial siege weapons are analogies borrowed from school desegregation, jury selection, and voting rights cases.

The variety of services potentially involved explains the enormous impact upon the fiscal policies and allocation practices of local government. Typical of such services—though by no means an exhaustive list—are the following:

---

[13] Holmes, "The Path of the Law," *Harvard Law Review*, 1897, pp. 457, 466-67.

paving, surfacing, and maintenance of streets; curb and gutter installation and maintenance; street-lighting and traffic control; recreational facilities; garbage collection; water for domestic consumption and fire protection; domestic utilities; fire-fighting services; and police patrols. Nearly all of these services were the object of an equalization complaint in Shaw, Mississippi.[14]

The nearly total segregation of the residential neighborhoods in Shaw was advanced as the ultimate explanation for the gross disparities in the level of services rendered to black and white citizens. A class action brought on behalf of the black residents presented the United States District Court for the Northern District of Mississippi with a statistical profile of both the output quality and input quantity of governmentally provided services, ranging from street paving and lighting to sanitary and storm sewers. The court was invited to draw the inference that *racial discrimination*, forbidden under the guarantees of the Equal Protection Clause of the Fourteenth Amendment, was the *prima facie* rationale for this state and condition. That court refused to enter such a finding, and in so doing posited its noninterventionist posture on the deference that courts have historically borne toward the elected officials of local government, and to the difficulties to be encountered were a judge to substitute his concept of what would be "ideal" levels of service for the political decisions reached through a majoritarian process.

The trial court's decision in *Hawkins v. Town of Shaw* indicated a strong judicial hesitation that has greeted attempts at law reform in service equalization suits. Having only recently discharged the burden of desegregating public accommodations and educational institutions, southern judges were understandably reluctant to engage problems that were vexing at the local level and staggering when projected to an eventual confrontation with the discriminatory practices of nearly every city and town throughout the land. On January 28, 1971, the United States Court of Appeals reversed the decision in *Hawkins*, and the retreating spirit of the judiciary as well.

In granting the plaintiffs' prayer for a declaratory judgment, the Court of Appeals declared: "The Town of Shaw, indeed any town, is not immune to the mandates of the Constitution. . . . 'A city, town or county may no more deny the equal protection of the laws than it may abridge freedom of speech, establish an official religion, arrest without probable cause, or deny the due process of law.' " With regard to formulating a remedy, the Court bypassed the arguments regarding the supposed inability of the judiciary to make resource-allocation decisions. "In concluding that an equal protection violation has occurred, we have not, of course, been guided by a statutory set of standards or regulations clearly defining how many paved streets or what kind of sewerage system a town like Shaw should have. We have, however, been

[14]*Hawkins v. Town of Shaw*, 303 F.Supp. 1162 (N.D. Miss. 1969).

able to utilize what we consider a most reliable yardstick—namely, the quality and quantity of municipal services provided in the white area of town."[15]

However it is ultimately decided by the Court of Appeals, the *Hawkins* case is still subject to an appeal to the United States Supreme Court. Affirmance by the Supreme Court would determine the legal questions on a national level. In the meantime, the direct precedent value of *Hawkins* has been to fashion an equalization mandate which can be enforced against any community within the multistate federal judicial circuit which has practiced racial discrimination in the distribution of municipal services.[16] In all other jurisdictions, the case presents a persuasive, though not formally binding, precedent. A victory of this magnitude is naturally encouraging, and the rapid dissemination of Judge Tuttle's opinion, coupled with national press publicity from New York to Los Angeles, is certain to stimulate local "replays."

The most immediate impact of the decision about Shaw, Mississippi, may be in California. In San Francisco, the city government already finds itself the target of a suit brought on behalf of the residents of Chinatown, complaining of the comparative lack in that most densely populated neighborhood of recreational facilities for either the young or the very old.[17] Attorneys for the San Francisco Neighborhood Legal Assistance Foundation complain that in a city which annually spends nearly $15 million for parks and recreation, less than $100,000 is allocated to this center of greatest population. The value judgments that will ultimately determine the outcome of this controversy will not be exclusively, nor even mainly, "local." The impact of a delta Mississippi village and a federal appellate tribunal in New Orleans will be vital, if not decisive, given the peculiar nature of the forum. As he journeys from the security of "city hall" to the dock of the United States District Court for the Northern District of California, the Mayor of San Francisco can attest that the dynamics of unanticipated change can overtake the institutions of metropolitan government with alarming force—and speed.

---

[15]The landmark nature of the *Hawkins* decision has prompted a motion by attorneys for the municipality that the judgment of the three-judge panel of the Court of Appeals be reviewed by the entire court. On May 24, 1971, the full court granted this petition for a rehearing *en banc*. The result of this review could be affirmance, modification, or reversal of the panel decision entered on January 28, 1971.

[16]The states included within the Fifth Circuit are: Texas, Louisiana, Mississippi, Alabama, Georgia, and Florida. The United States Court of Appeal has its headquarters in New Orleans, Louisiana.

[17]*Woo v. Alioto*, Civ. No. 52100 ACW (N.D. Cal., filed November 16, 1970).

For Product Safety Concerns and Information please contact our EU
representative GPSR@taylorandfrancis.com
Taylor & Francis Verlag GmbH, Kaufingerstraße 24, 80331 München, Germany